PIERCING THE SURFACE
X-RAYS OF NATURE

CARLO AND STEFANO GRECO

PIERCING THE SURFACE
X-RAYS OF NATURE

HARRY N. ABRAMS, INC., PUBLISHERS, NEW YORK

The photographs and radiographic images are by Carlo Greco

The authors would like to thank Carlo Loreto,
Guido Politi, Nunziato Ricciardi, Casimiro Nicolosi,
Gaetano Finocchiaro, and Giovanni Aleo

LIBRARY OF CONGRESS CATALOGING-IN-PUBLICATION DATA

Greco, Carlo.
    Piercing the surface.

    Translation of: Natura ai raggi X.
    1. Radiography. 2. Photography, Biological.
I. Greco, Stefano. II. Title.
TR750.G7413   1987   574'.028 87–1426
ISBN 0–8109–1495–6

Times Mirror Books

Printed and bound in Italy by SAGDOS S.p.A., Milan

"X Rays of Nature" might sound like the subtitle for a mainly scientific book, but this is not the case: our radiological study of the living world around us does reveal some features having scientific importance, but others that are visual, with artistic, poetic, or philosophic overtones, according to one's mood, outlook, or cultural background. The naturalist will value the scientific aspects of an image; the imaginative observer will see the fantastic; and the artist will appreciate the beauty of the shapes, and give symbolic and figurative meaning to images that speak to the viewer's eye and spirit. It was our recognition of this wide range of interesting responses that led us to embark on this book.

Many people might question the purpose of making a radiological survey of objects that have already been amply studied by sectional and dissectional examination. To this we have at least two answers: first, it is interesting to study the whole object without splitting it up mechanically into parts; second, there is always the hope of obtaining a stimulating, fresh image. Success often crowned this hope unexpectedly; some objects' radiological pictures proved to have more charm than their external appearance.

Radiology allows us to go beyond the limits of human perception, deeper and farther, and it can be intriguing or moving to discover the beauty and harmony that lie beyond these barriers. Radiography is a modern answer to the curiosity that natural phenomena have always aroused in us as we strive to penetrate the world around us; from time immemorial we have tried to fathom hidden secrets, and the most subtle minds have been passionately engaged in observing the living world. In this quest, the work of naturalists must not be underestimated, from Pliny the Elder (who risked and then lost his life in observing the eruption of

Mount Vesuvius) to the great Carolus Linnaeus, the supreme systematist of the natural order; nor should we overlook the contribution of the great scientists and philosophers, from Aristotle to Lucretius, and Leonardo da Vinci; and always the search for reality, transfigured but none the less true, by artists and poets in every time. Perhaps only they can record and capture the breathing and trembling of the living world—from Virgil, who contemplated the murmuring of the bees among the hedges and willows, to William Wordsworth, whose heart danced with the daffodils.

All sensitive persons have approached with trepidation this vast "book" that opens before our eyes, this spectacle that daily offers fresh food for the senses. The reality observed through radiology may well appear unusual, but it is no less real than that normally perceived. If this technological advance enables us to read and enjoy new pages in the "book" which our forebears were unaware of, surely it is to be used for exploring regions hitherto obscure.

We should like to view radiology not as a technological means of stripping the veil of enchantment from the living world, but as a magic eye that reveals and highlights exciting aspects of the world, and not on the aesthetic plane alone. We hope that others will be stimulated by the fascination of X-ray images to penetrate ever deeper into the mysterious world of life, still so full of secrets.

In conclusion, we should like to thank those who have contributed to this book, in particular the Italian firm of Gilardoni, makers of Soft-Gil equipment that enabled us to take radiographs of even the thinnest specimens, and the World Wildlife Fund, which provided many of these specimens.

CONTENTS

INTRODUCTION

Evolutionary thinking must inevitably be taken into account when writing a book on the subject of nature. There can be no doubt that, in the light of recent evolutionary findings, all living organisms are connected by an uninterrupted biological thread. Linnaeus, the first real naturalist of modern times, expressed this concept with the aphorism *"natura non facit saltum"* ("there are no leaps in nature"). But the various forms of life do not merely follow one another along a linear sequence, leading from simpler forms to more evolved ones. In the course of evolution, diversification has given rise to numerous branches off the main biological "trunk," thereby producing the great variety of living organisms. It is possible to represent the phylogenic sequence of the life forms as a treelike diagram whose main branches are the fundamental divisions within the kingdom of the phyla.

To keep this book from appearing overly scientific, it was decided to give each chapter a title familiar to everyone. At the same time, these ordinary names—flowers, shells, sea urchins and starfish—correspond to well-defined taxonomic groups of organisms, and the scientific nomenclature of the phylum or class is given in the subtitles.

Most of the pictures in the book are of the animal world,

fascinating for its vast variety. Radiological investigation of such organisms produces images having undoubted scientific as well as visual interest.

The first chapter, however, is devoted to the plant world, of which flowers are the only aspect considered. Flowers are the expression of the most evolved plants, the phanerogams. It was decided that the first chapter should be on flowers for a precise ecological reason: without the existence of plants, there would be no form of animal life to be treated.

Although we chose a phylogenetic baseline, practical and technical limitations kept us from dealing with all aspects of "life." Our main purpose was to produce images interesting from several viewpoints, in particular the aesthetic one; beautiful and stimulating pictures were selected over those less beautiful, though perhaps having richer scientific significance.

For consistency with such an aesthetic choice, both the photographic and radiological images are shown for those organisms in which both are pleasing. Of those organisms producing, often surprisingly, a radiological image more interesting than any photograph of their exterior form, only the radiograph was chosen.

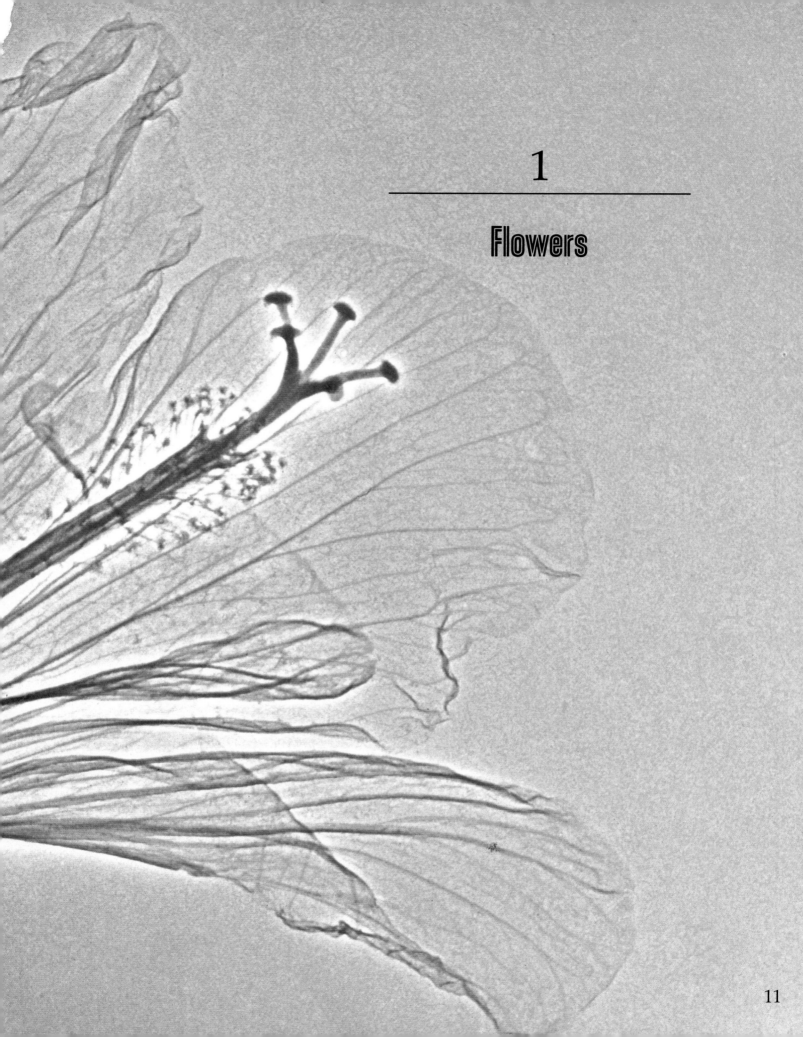

Flowers, with their delicate grace and harmonious shapes, are the most highly evolved and sophisticated expression of plant life. Their charm has the precise, built-in objective of attracting particular insects, and they achieve this by their colors, shapes, perfumes, and nectar. Thanks to the insects' unconscious contribution, pollination takes place, which is absolutely essential for reproduction of the plant species. In short, the flower is the reproductive organ of the more evolved plants, the phanerogams, and the energy spent by these plants to produce the flower with all its attributes is in the interest of the plants' posterity. In carrying out this task with such splendor, the flower arouses intense emotion.

Although its radiological image does not show those attributes of the flower that make it so beautiful in real life, it reflects its own delicacy and transparency. Moreover, one can appreciate the whole flower, since the external and internal structures are simultaneously exposed. By looking through the petals, which are like thin veils, radiology permits one to see the flowers' internal organs, the pistil and stamens. In full bloom, the flower is revealed at the height of its beauty yet in all its intimacy; the radiological image adds little to what close external observation yields, but it allows admiration from a new viewpoint. Fresh images are created, attractive in their purity on a visual level.

Yet before the petals unfold, the flower jealously guards its intimate structures. These are revealed by radiological examination; enclosed in a restricted space, the individual parts are seen in close contact with each other.

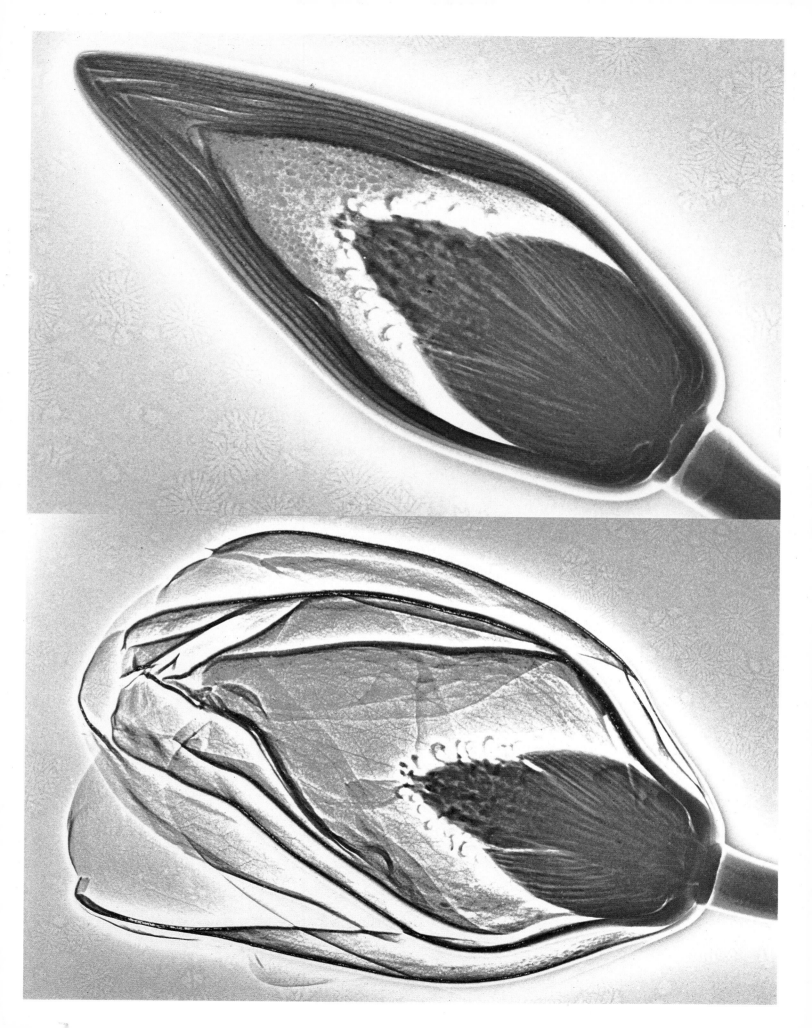

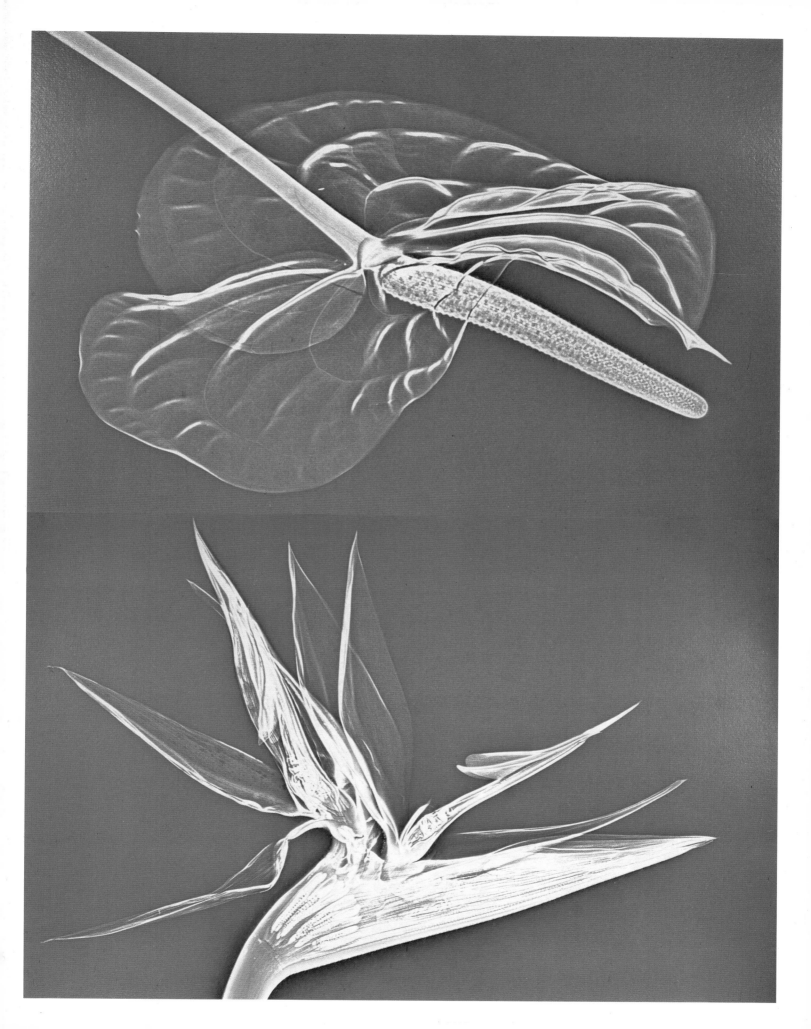

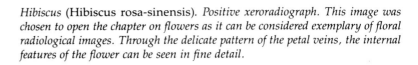

Hibiscus (Hibiscus rosa-sinensis). *Positive xeroradiograph. This image was chosen to open the chapter on flowers as it can be considered exemplary of floral radiological images. Through the delicate pattern of the petal veins, the internal features of the flower can be seen in fine detail.*

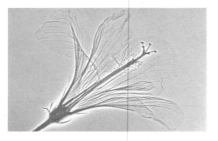

Fuchsia (Fuchsia). *Photographs and radiographs of bud and flower. The chromatic harmony of the bud and flower of the fuchsia is complemented by the charming patterns of their radiological design.*

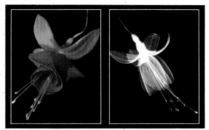

Day Lily (Hemerocallis). *Photographs and radiographs of the flower in two different projections. X rays reveal the remarkable pattern of fibers in the petals, which give visual variety to the radiological images according to the projection adopted.*

Frangipani (Plumeria). *Photograph and radiograph. The photographic image renders only part of the delicacy of this highly perfumed blossom and it appears equally charming radiologically, though deprived of its tender colors. The outstanding feature of the radiograph is the delicate vascular system that nourishes the petals.*

Lily (Lilium). *Two photographs and two positive xeroradiographs of the flower in different projections.*

Hibiscus (Hibiscus). *Photographs of two varieties of hibiscus, and radiographs in two projections. These radiographs of hibiscus flowers, although less detailed than the xeroradiographic image (see above), highlight just as effectively the splendid pattern of veins in the petals, doubtless the special radiological feature of this flower. Because the concentration of water, a radio-opaque element, is greater in the veins than in the surrounding tissue of the petal, a good radiological contrast is created.*

Rose (Rosa). *Photograph and radiograph. In the radiological image of a rose from the side we can see through the numerous petals into the reproductive structures of the flower, which, indeed, are not highly developed.*

Wax Plant (Hoya carnosa). *Photograph and radiograph. Hoya flowers are gathered in beautiful inflorescences consisting of numerous star-shaped, waxy blossoms, highly perfumed. From the radiograph of one of these there emerges a harmonious geometrical image.*

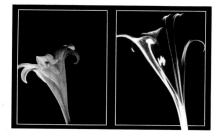

Regal Lily (Lilium regale). *Photograph and radiograph. The radiological appearance of this white flower maintains most of its great delicacy, revealing also its innermost structures, gracefully displayed inside the calyx.*

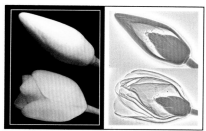

Southern Magnolia (Magnolia grandiflora). *Photographs and positive xeroradiographs of bud and flower. The rounded, elegant arrangement of the rich petals in the radiological image of the opened flower, below, is in contrast with the motionless layers of the bud petals, above. These petals remain separate from the reproductive organs, though all are so intimately tucked up together.*

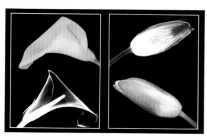

Left: Calla Lily (Zantedeschia aethiopica). *Photograph and radiograph. Calla lily flowers are gathered in the inflorescence of a yellow spike, fully visible in the radiological image, protected by a snow-white spathe which reveals a considerable fibrous structure.*

Right: Water Lily (Nymphaea alba). *Photograph and radiograph. The water lily opens with the sunlight, and closes with the night. The radiological image of the closed flower allows us to see the arrangement of the petals and the more intimate structures, all protected by the fleshy external sepals.*

Above: Anthurium (Anthurium). *Photograph and negative xeroradiograph. The presence of more fleshy parts on the red spathe, the modified leaf that crowns the inflorescence, creates radiologically an almost ethereal picture through its intertwining lines.*

Below: Bird-of-Paradise (Strelitzia reginae). *Photograph and negative xeroradiograph. With X rays to penetrate the spathe of the bird-of-paradise, we can observe the inner structures that produce its beautiful inflorescence.*

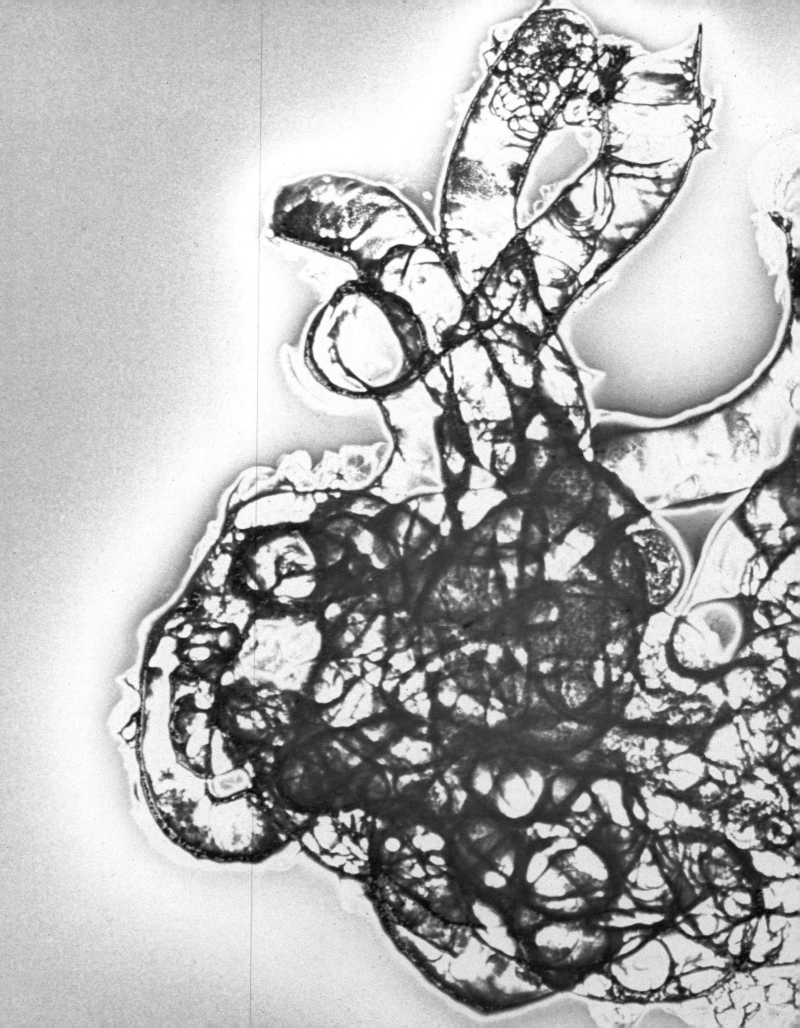

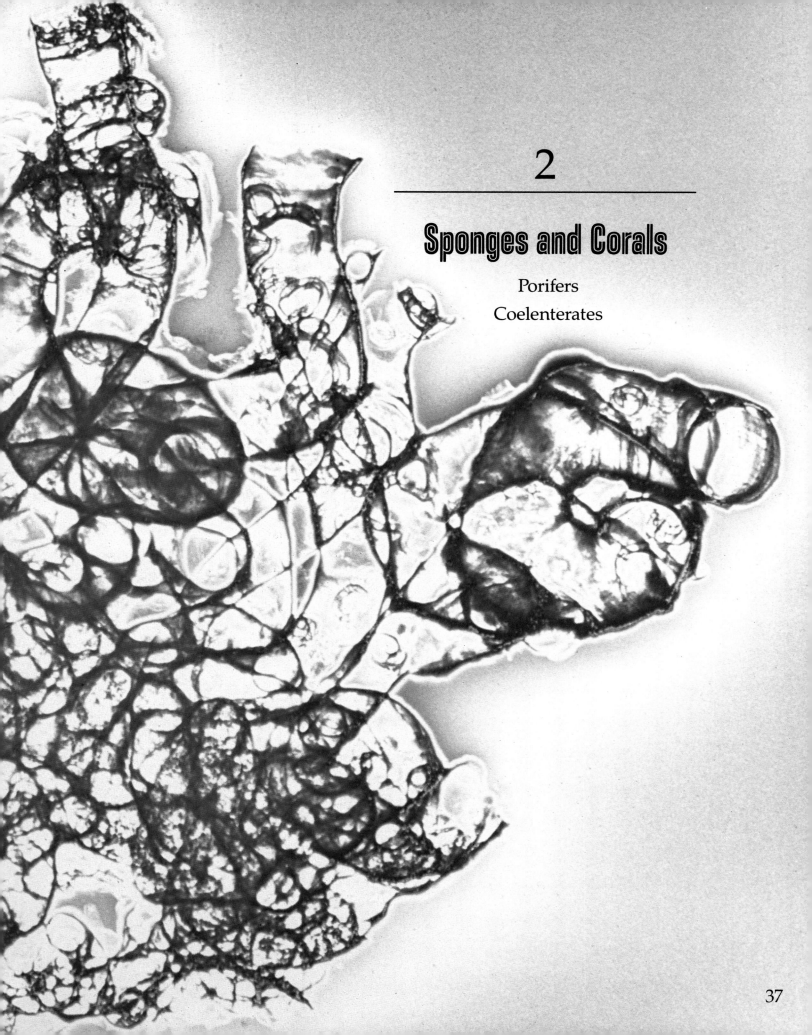

2

Sponges and Corals

Porifers
Coelenterates

Animal life began with the appearance of protozoa, one-celled organisms; these gave rise to multicellular forms of animal life, the metazoa, through stages of cellular aggregation and differentiation.

Sponges, the *Porifera*, are the most elementary group of metazoa. Despite their simplicity, sponges already synthesize the fundamental features of an animal organism, in which various types of cells, each performing a specific function, contribute to the health of the whole organism (were this not so, sponges would be regarded as colonies of unicellular organisms). Within their bodies can be observed different types of cells, one of these being concerned with the secretion of calcareous or limy materials in the form of hard spicules, which form a kind of skeleton. Such a skeletal structure in sponges can be revealed radiologically.

Sponges evoke great interest as one of the most primitive stages of the evolution of metazoa, but it is generally believed that they are a sterile branch of the evolutionary tree, not part of the evolutionary line that goes on to higher animal forms.

One step up, we come across the phylum, *Coelenterata*, "hollow-bodied," which includes groups of organisms familiar to everyone, such as corals, jellyfish, and sea anemones. Even though coelenterates might appear in some ways little more evolved than sponges, they exhibit a more complex organization, which, even in its simplicity, reflects that of all higher animals. Coelenterates have a mouth, a gut (absent in sponges), and the rudiments of a nervous system.

The class *Anthozoa*, comprising organisms such as corals and polyps, is without doubt the most interesting of the *Coelenterata*. Radiological investigation of the skeletons of these organisms emphasizes the extraordinary beauty of their structure, characterized by filled and empty spaces; these, with their shapes and distribution, give radiological images of great interest for their formal and spatial qualities, open to imaginative interpretations by the observer.

To complete our radiological overview of forms of animal life in this chapter, a radiological image of a marine polychete has been included. These wormlike organisms belong to the phylum *Annelida* and live inside long, calcareous tubes. Annelids are a further step along the evolutionary path as compared to coelenterates, having a more complex nervous system that controls every segment of the animal. Probably it is through the ancestral forms of annelids that the evolutionary thread continues uninterrupted.

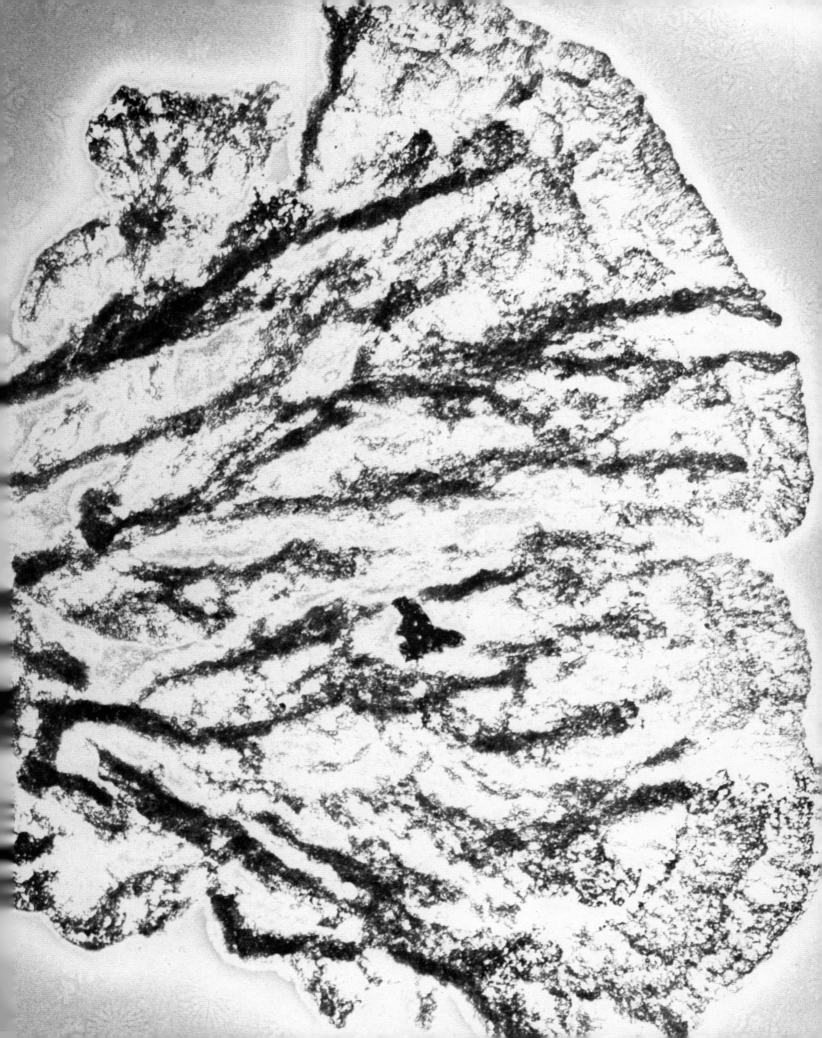

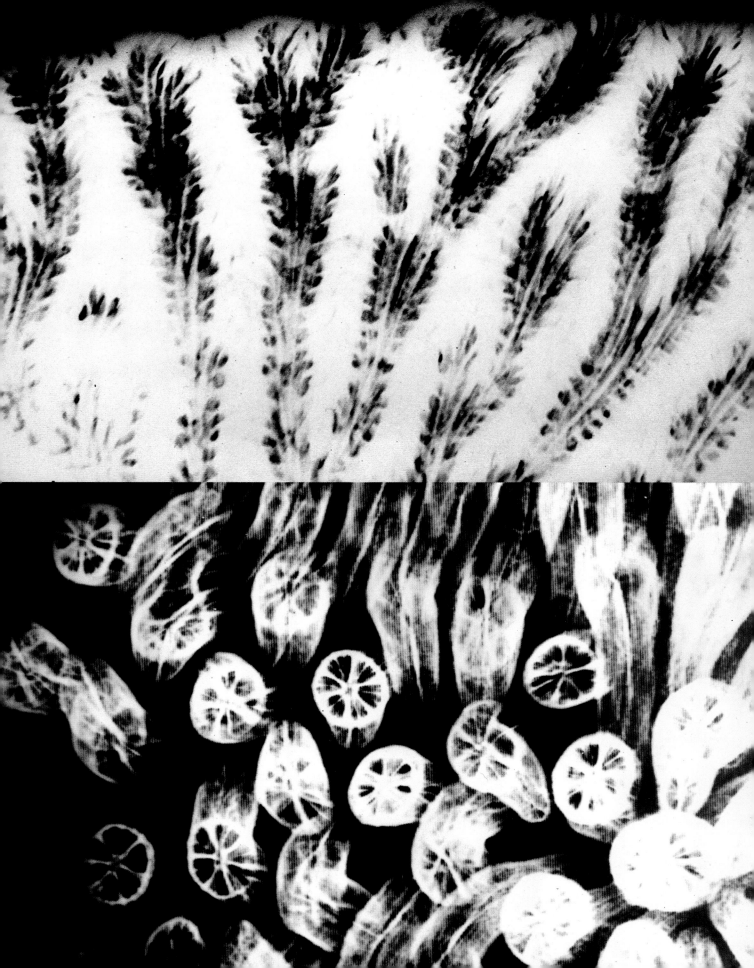

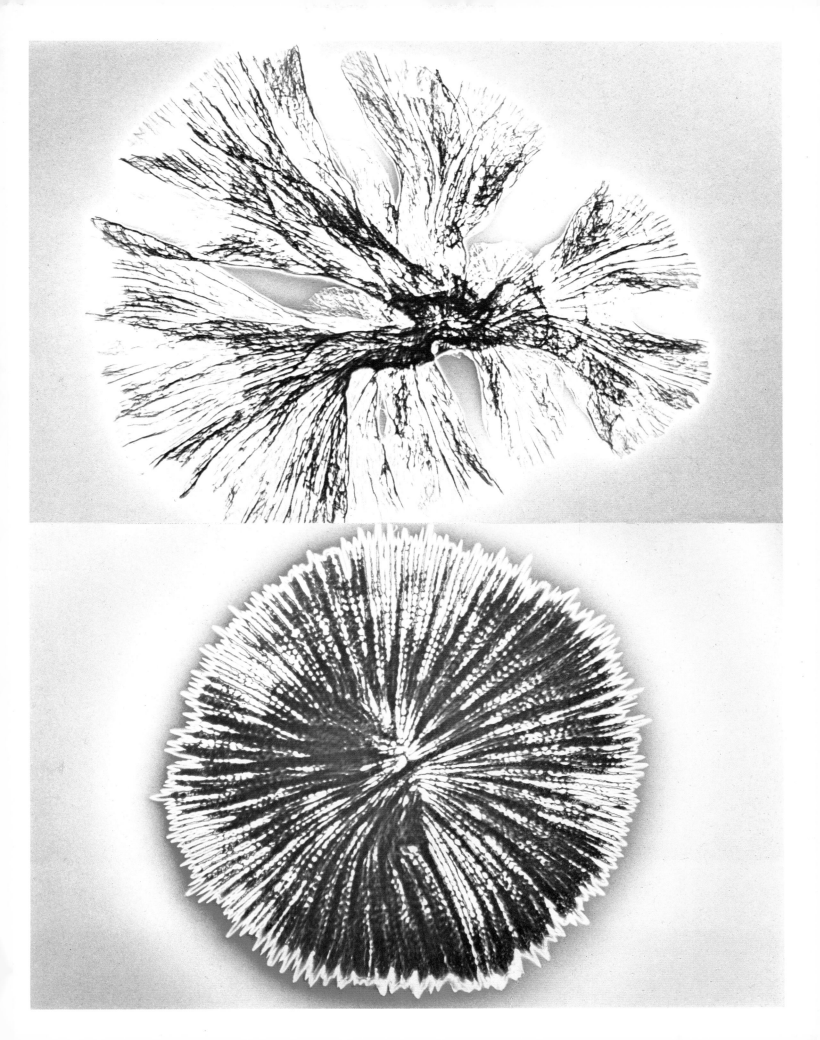

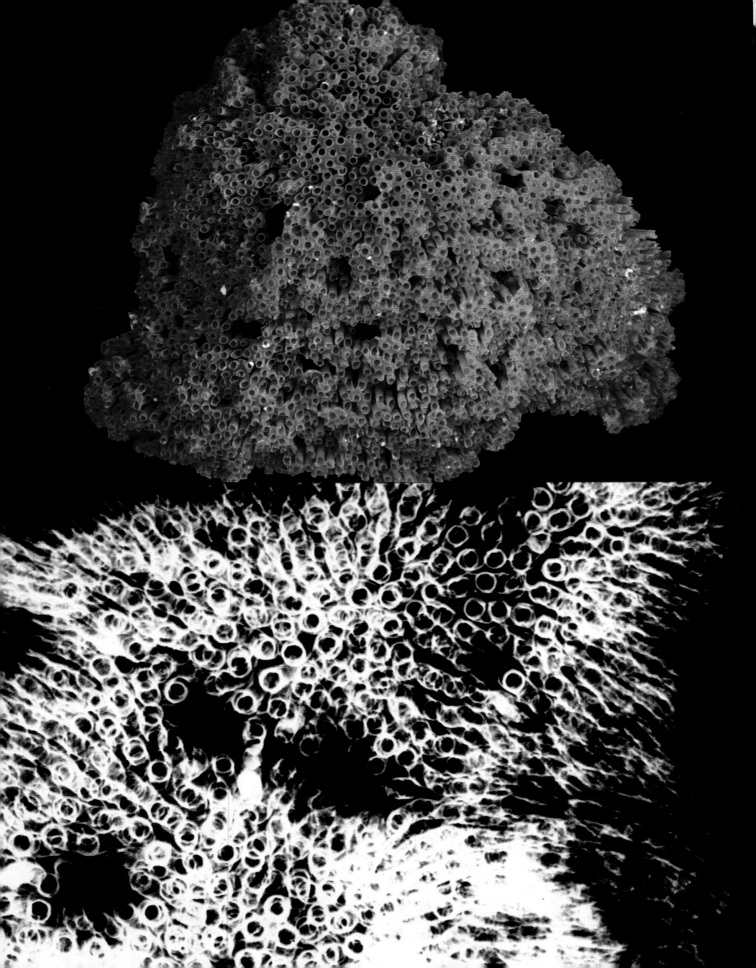

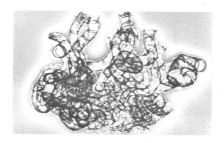

Tube Worms (Protula sp.). Positive xeroradiograph. The genus Protula *comprises several species of marine polychetes or tube worms which live inside long, contorted limy tubes variously intertwined. These organisms belong to the phylum* Annelida *and are considered more evolved than sponges and corals; they have been included here for the amazing fascination of their radiological image.*

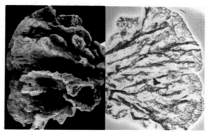

Bath Sponge (Euspongia officinalis). Photograph and positive xeroradiograph. Despite the well-known softness of sponges, X rays demonstrate the presence of a limy or calcareous bearing structure.

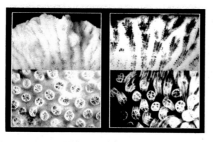

Above: Pavona sp. *Photographic and radiographic details of the skeleton of a coral colony. X rays show that the thinner parts of the calcareous skeleton have real solidity.*

Below: Galaxia sp. *Photographic and radiographic details of the skeleton of a coral colony. With X rays, the skeleton of each polyp can be isolated from the limy matrix within which it belongs.*

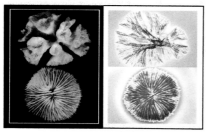

Above: Eusmilia sp. *Photograph and xeroradiograph of the skeleton of a coral colony. We can observe radiologically the partitions or septa that make up the length of the skeleton.*

Below: Mushroom Coral (Fungia sp.). Photograph and positive xeroradiograph of this species, which resembles the gills beneath a mushroom cap.

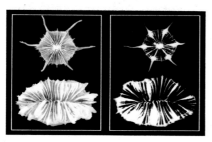

Above: Dendrophyllia profundorum. *Photograph and radiograph of the skeleton. This small, relatively solitary coral, living at great depths, does not form reefs.*

Below: Lobophylla sp. *Photograph and radiograph of the skeleton.*

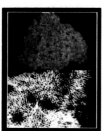

Organ-pipe Coral: (Tubipora musica). Photograph of the skeleton of a coral colony; radiographic detail, much enlarged. The polyps that make up the colony live inside slender, limy tubes which create a structure suggesting the pipes of an organ.

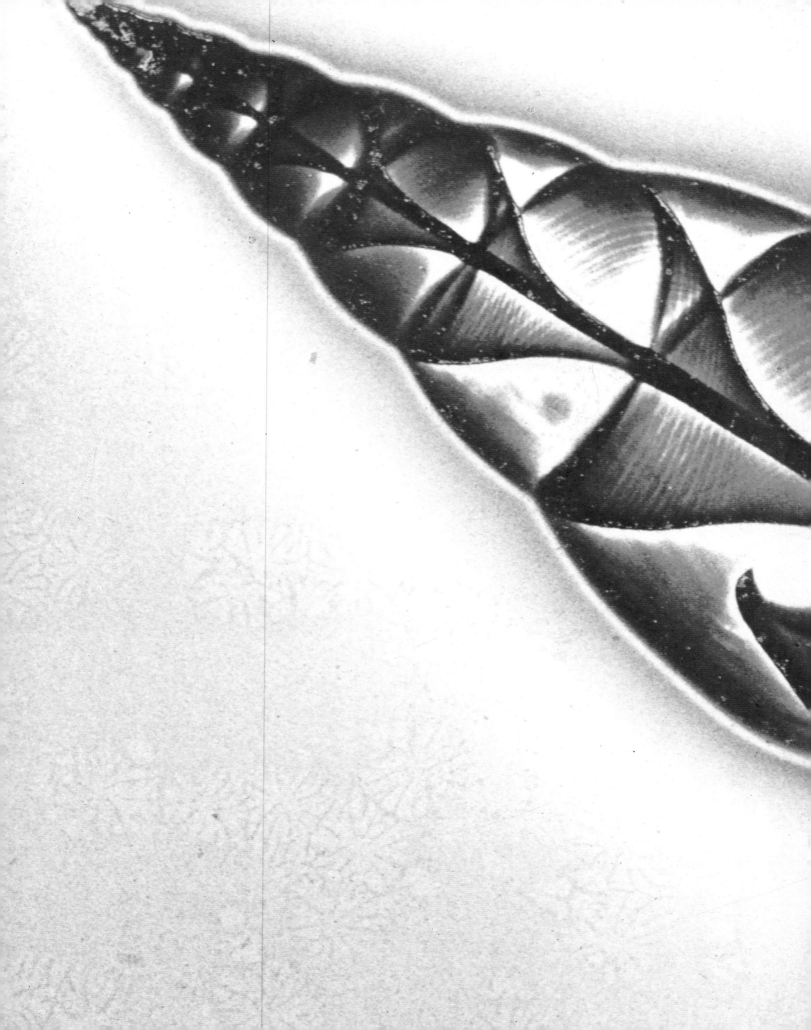

3

Shells

Mollusks

Behind the fascinating world of seashells lies one of the most numerous and diversified groups of invertebrates, the phylum *Mollusca*.

Mollusks probably originated from the ancestral forms of annelids (see Ch. 2), and they owe their extraordinary biological success to the acquisition, very early during their evolution, of the mantle, a thin cutaneous layer covering the whole body of the animal. The mantle is capable of secreting a limy material at one or more points, producing a hard shell; the animal can withdraw into this in case of danger.

In their structure, shells vary considerably among the many classes of mollusks: they can be single, bivalve, reduced to a mere remnant, or, indeed, completely absent. On the basis of shell morphology and anatomical and physiological differences, six classes are distinguished within the phylum: most important are gastropods (univalves), bivalves, and cephalopods.

Gastropods possess shells of great beauty and structural harmony, characterized by their spiral structure, the result of a peculiar and unique phenomenon known as "torsion." It is believed that the early gastropods did not have a conical shell but rather a flat one, the spires (whorls) interwrapped within a single plane. Successively gastropod shells exhibited an asymmetrical wrapping of the spires which grew bigger along the horizontal and vertical axes, giving rise to the typical conical, spiral shape. This basic motif has been "interpreted" by gastropods into a vast array of solutions, producing fascinating shapes, ornaments, and colors that have always evoked deep emotions in us.

Radiological images of gastropodal shells go beyond the mere revelation of internal structures. We are led to explore mazes of spires, and to penetrate a whole new world of fantasy, baffled by the harmony and perfection of nature's extraordinary architectonic solutions.

Totally different externally, but no less interesting, are the shells of the *Bivalvia* mollusks. As the name implies, these shells are formed of two hinged valves which can be closed in case of danger. Their radiological images do not highlight structures as complex as the gastropods', but they bring out features that are visible on the exterior yet often underestimated. A subtle form of elegance is seen to lie behind their great simplicity. A good example of this is the radiological image of a scallop shell, *Pecten jacobeus*, suggesting a luminous fan.

The cephalopods are the highest and most active class of the mollusks, with an extraordinary evolutionary history. Nowadays this class is limited to a few genera, of which only one, *Nautilus*, maintains its external shell, all the others' having been either reduced or lost altogether. *Nautilus* is regarded as a living fossil, the only survivor of a vast order of cephalopods to which ammonites once belonged. Once extremely widespread, ammonites suddenly became extinct at the end of the Cretaceous period.

The radiological image of the *Nautilus* shell is of the greatest structural, visual, and biological interest, for thereby, without cutting into it, we can admire its internal structure where the developmental stages of the animal are visible. While growing, the animal creates successive chambers, separated by septa, and lives in the last and largest of these.

Thus radiology has the great merit of affording us a new approach to the world of shells, defining visually their structure and formal features, their transparency and fluorescence, and the mysterious message of their internal ornamental designs.

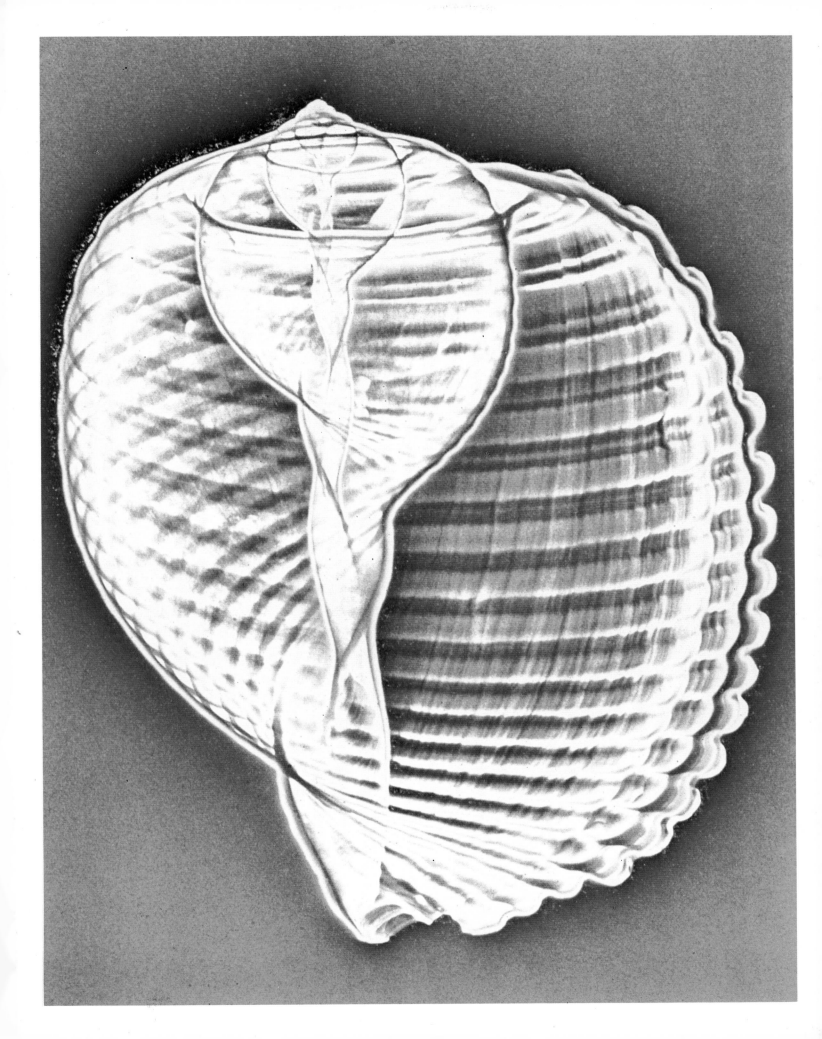

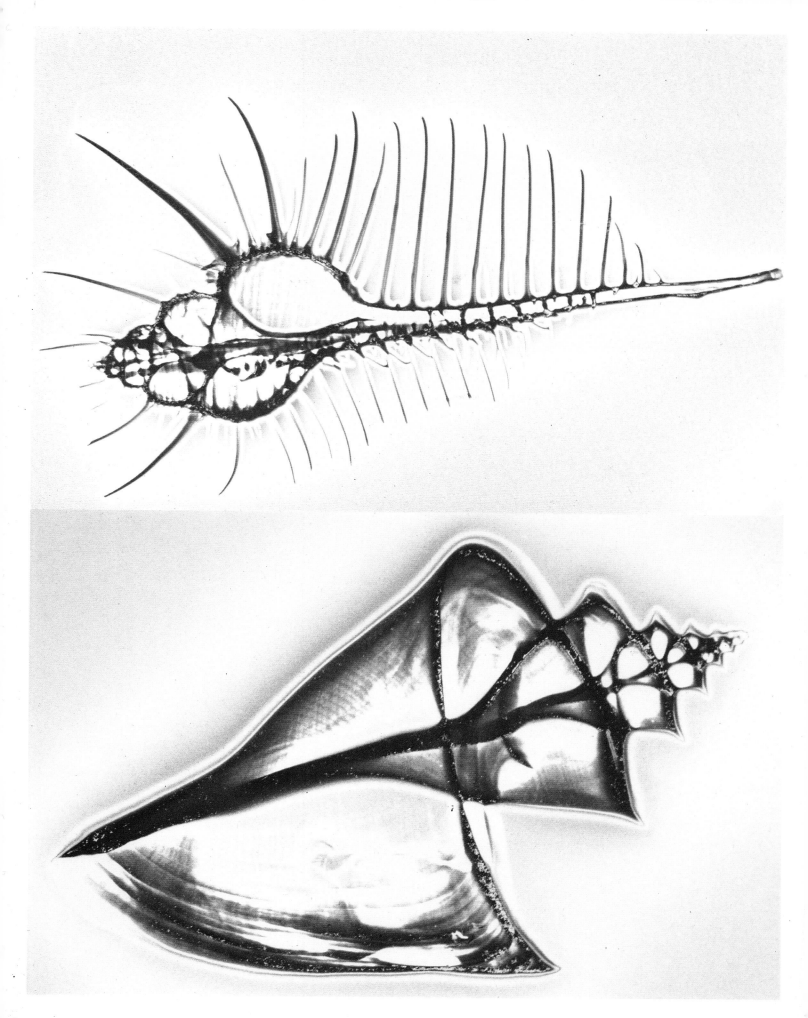

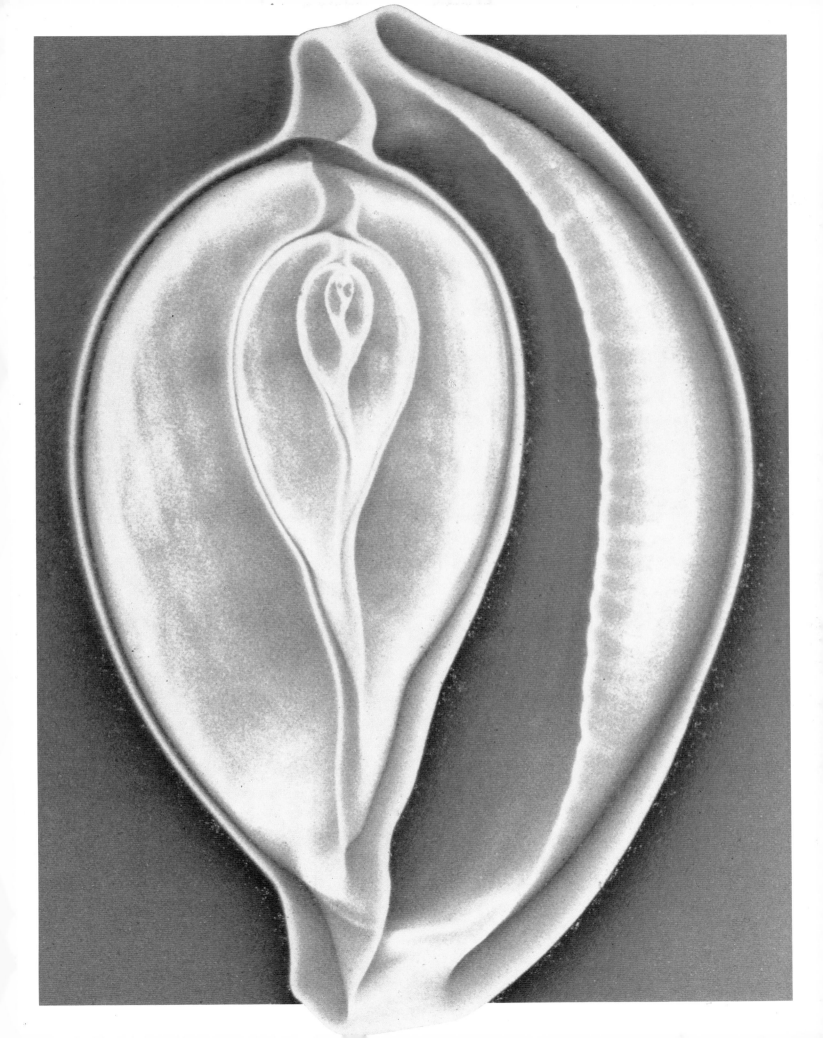

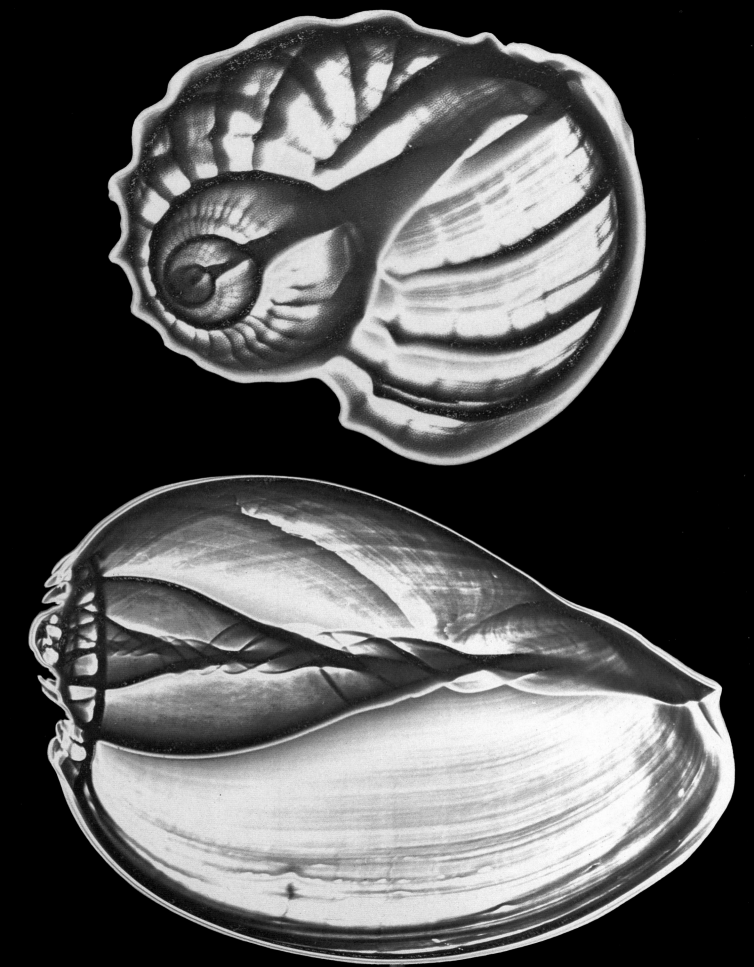

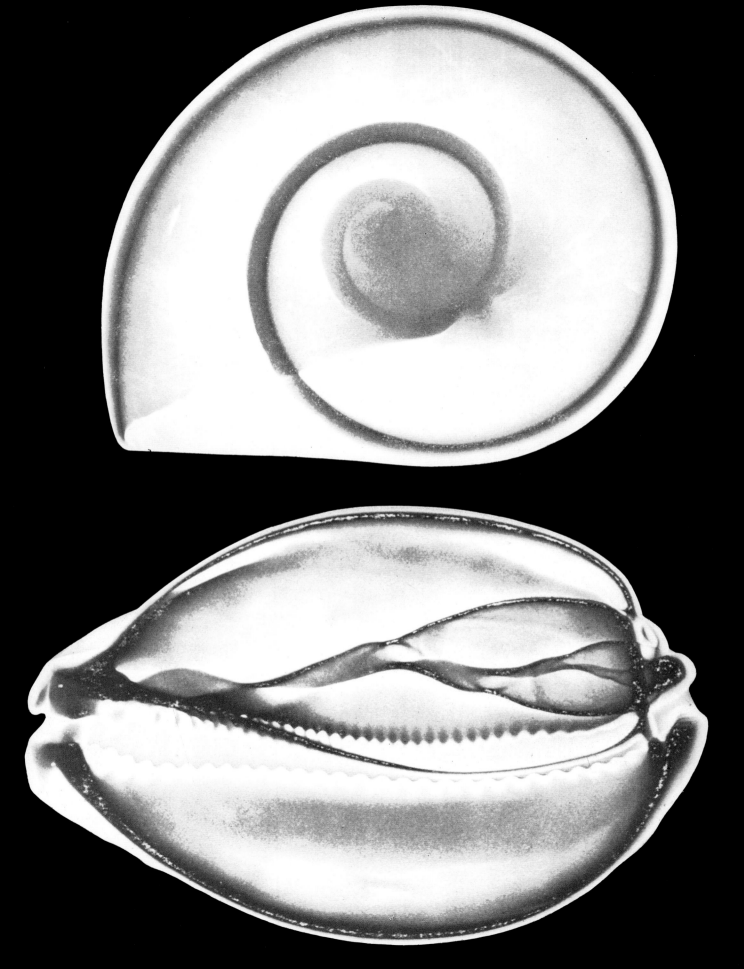

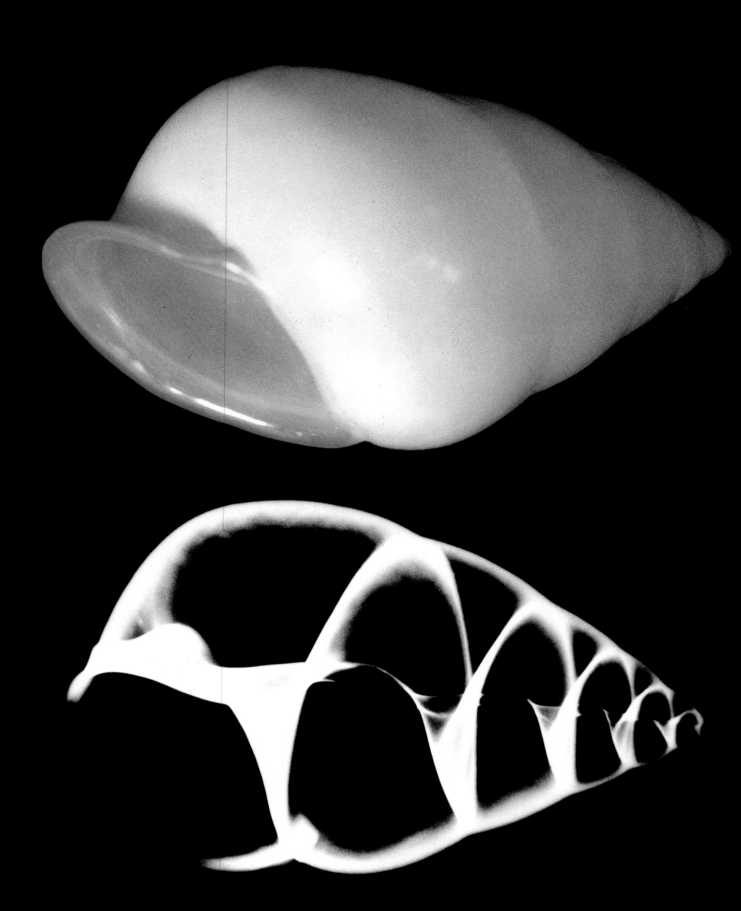

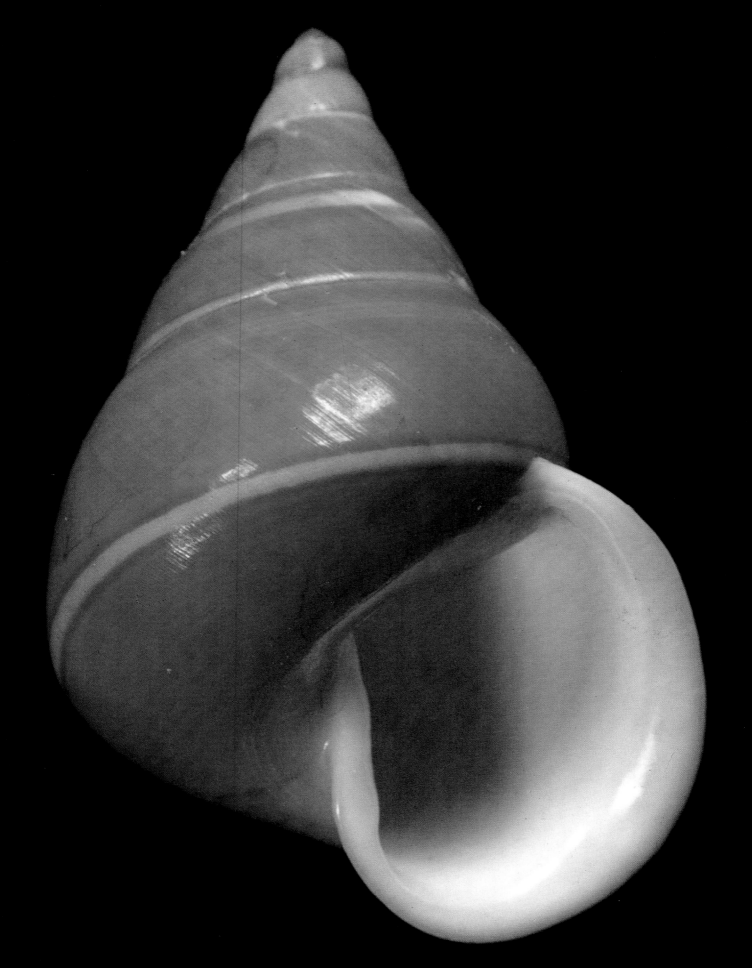

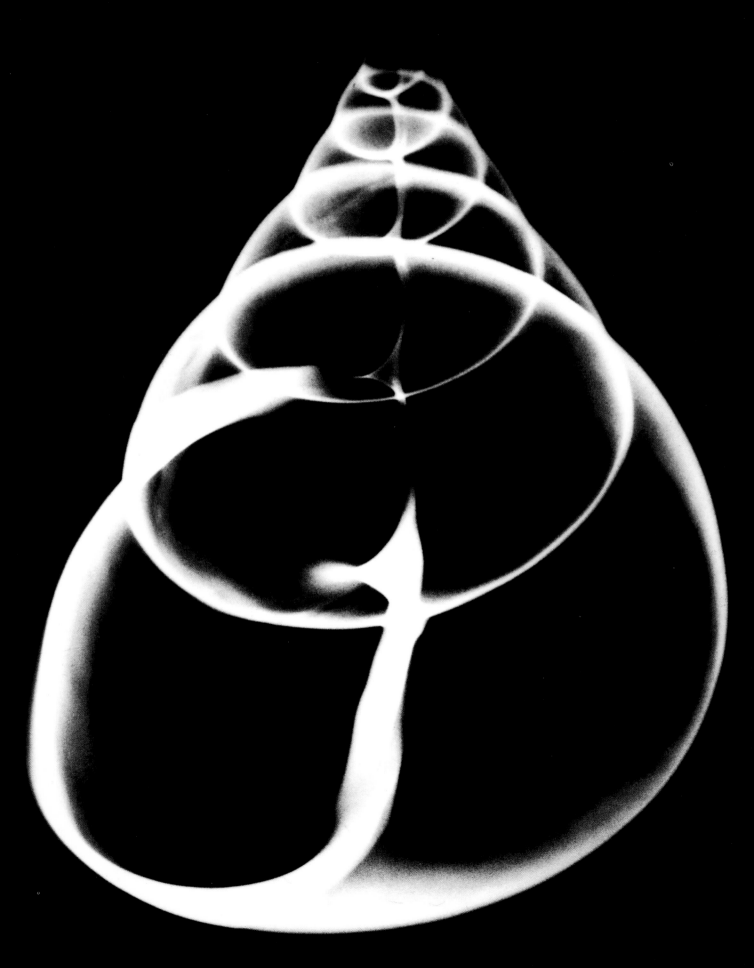

Martini's Tibia (Tibia martinii). Positive xeroradiograph. The thin walls of this shell allow a picture of its internal structure in its smallest details, keeping the three-dimensionality of the spiral almost unaltered.

Tun Shell (Tonna galea). Left: Two photographs of the shell, and two radiographs in different projections. Right: Negative xeroradiograph. This fragile, ribbed tun shell, very common in the Mediterranean Sea, hides exciting radiological aspects behind its simple but elegant exterior. The radiological images vary enormously according to the projection.

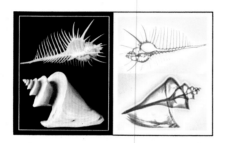

Above: Murex Shell (Murex pecten). Photograph and positive xeroradiograph. The appeal of this murex shell is certainly due to the exotic beauty of its ornamentation. Three rows of spines extend along the short spiral and down the long siphonal canal.

Below: Miraculous Thatcher Shell (Thatcheria mirabilis). Photograph and positive xeroradiograph. The beauty of this shell lies in the essential simplicity of its lines. Its internal design is equally simple and also has a spatial movement both ample and delicate.

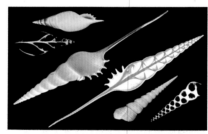

Above left: Martini's Tibia (Tibia martinii). Photograph and radiograph (and see xeroradiograph above). Center: Tibia fusus. Photograph and positive xeroradiograph. The extraordinary elegance of this tibia shell comes from the perfect balance of the length of the spiral with that of the delicate siphonal canal. In the radiological image it is interesting to observe that the siphonal canal is a continuation of the columella inside the spiral. Below right: Augur turritella (Turritella terebra). Photograph and radiograph.

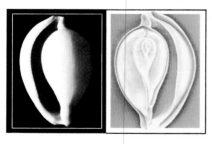

Ovula ovum. Photograph and negative xeroradiograph. Although these "egg shells" are similar to cowrie shells, they are less sought after. The internal structure of the Ovula shell is often more delicately formed than that of many cowries.

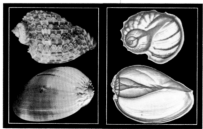

Above: Harp Shell (Harpa major). Photograph and positive xeroradiograph. Most characteristic of harp shells is their sturdy bright-colored ribbing. The radiological image that emerges, developing from the ribbed projections of the shell, is usually interesting.

Below: Melo melo. Photograph and positive xeroradiograph. It is interesting to see in the xeroradiographic image the "movement" enfolding the columella.

Left: Cone Shell (Conus marmoreus). Photograph and radiograph. The typical shape of cone shells comes from the concentric wrapping of the whorls around the vertical axis of the columella. This makes a superimposition of the whorls; to enrich the image, an oblique rather than a lateral radiological projection must be adopted.

Right: Triton Shell (Apollon perca). Photograph and radiograph.

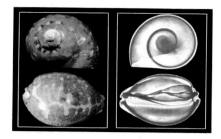

Above: Turban Shell (Turbo marmoratus). Photograph and positive xeroradiograph. From the pure lines of the radiological picture of a turban shell in an axial projection, the growth of gastropodal shells emerges clearly according to a logarithmic spiral, the distance between the walls increasing geometrically from the inside to the outside of the spiral.

Below: Map Cowrie (Cypraea mappa). Photograph and positive xeroradiograph. This array of transverse folds ornamenting the aperture of cowrie shells is very apparent in the xeroradiograph.

Papustyla xanthochila. Two photographic and (below) two radiographic images in different projections. This shell is one of the chromatic varieties belonging to the genus Papustyla, terrestrial snails of New Guinea. The radiological images show the columella of the shell to be hollow; this characteristic is deducible by external observation from the presence of an umbilicus, *a small cavity in the center of the base of the spire. The radiographic image at upper center, does not show this characteristic, being of a different species, Papuina pulcherrima.*

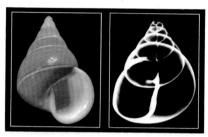

Papuina pulcherrima. Photograph and radiograph. The simplicity and subtle elegance of the forms in this radiological picture exemplify the finest of shell radiology.

Garden Snail (Helix sp.). Left: Photograph and positive xeroradiograph. Right: Radiographs of three different projections. The shell of a common snail "hides" lines of great harmony and elegance which can be modulated by the radiological projection that is adopted.

Left: Scallop Shell (Pecten jacobeus). Photograph and radiograph. The radiological image of this bivalve scallop shell, emphasizing its radial ribs, suggests a luminous fan.

Right: Brown Paper Nautilus (Argonauta hians). Photograph and radiograph. Paper nautiluses are a genus of cephalopods which lack an external shell. The females, however, secrete a very fragile shell and lay their eggs within it, where the young hatch.

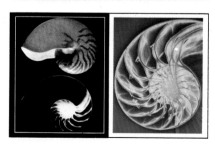

Chambered Nautilus (Nautilus pompilius). Left: Photograph and radiograph. Right: Negative xeroradiograph, detail, of center of shell. The radiological image of the nautilus shell has many features of scientific interest. One can see the chambers, separated by septa, that represent the developmental stages of the animal, which lives in the last and largest of these. Also visible is the siphonal canal connecting the successive gas-filled chambers with which the animal balances itself in the water.

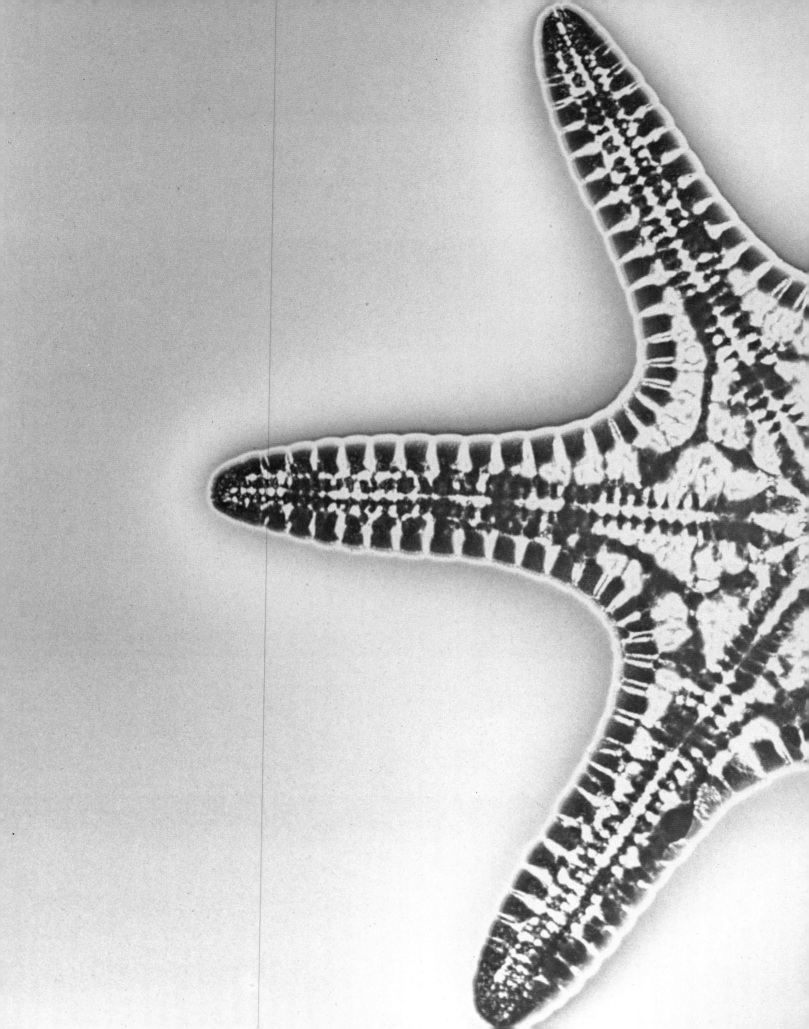

4

Sea Urchins and Starfish

Echinoderms

Sea urchins and starfish are the most representative animals of the phylum *Echinodermata,* a group of organisms whose origin is rather uncertain; from the early stages of its evolutionary history, it has exhibited some peculiar features not found elsewhere in the animal kingdom.

The most obvious characteristic of the echinoderms is their five-part or pentamerous symmetry, which has led to the hypothesis that originally the echinoderms were sessile, fixed to a base through a stalk, for the bilateral symmetry normal to more evolved organisms loses its meaning in such conditions. A sedentary organism tends to develop its body uniformly in every point, creating a radial symmetry, and a substantial class of sessile echinoderms, the sea lilies, still exists today. Sea urchins and starfish have abandoned a sedentary life and are capable of moving freely, yet they maintain a radial symmetry, having no apparent meaning, that is explained by the evolutionary history of the phylum.

Also typical of the phylum is the presence of an internal skeleton, the dermaskeleton, made up of small calcareous plates either fused together or articulated jointly with one another. This type of skeleton, surrounded by living tissues, differs from the shells of mollusks, which are external to the animal tissues. The radiological investigation of these echinoderms allows us to observe the arrangement of the little plates and to distinguish the two most important classes of echinoderms, the echinoids or sea urchins, and the asteroids or starfish.

In the echinoids, the calcareous plates are fused together and constitute a continuous structure; within it are five radial elements, known as ambulacra. These are characterized by a series of small holes through which pass the tube feet, little pedicels connected to the "water-vascular" system, a structure peculiar to echinoderms. Originally the tube feet probably had only a respiratory function, permitting the passage of water in and out of the animal's "water-vascular" system, but in modern echinoderms they have taken on a locomotive function, enabling the animal to move around freely on the sea bottom.

Sea urchins feed on algal matter, which they take in through their mouth at the center. This mouth is a complex chewing apparatus consisting of five sharp, triangular teeth; it is known as Aristotle's lantern, because the great Greek philosopher first described the organ. It can be observed radiologically by examining a live sea urchin from the side.

Sand dollars belong to a category of echinoids that are flat and irregular in shape, and known as irregular echinoids (regular echinoids have a spherical shape). Radiological images of these organisms show in the center a chewing mechanism that is comparable to the Aristotle's lantern of regular echinoids. Calcareous "buttresses" that join its lower and upper faces are revealed radiologically inside the test (outer shell) of these animals. These buttress structures produce an intriguing pattern in which the spaces occupied by the internal organs of the animal can be recognized.

In starfish, or asteroids, the calcareous plates that form the skeleton are not fused together into a test but are articulated to give flexibility to the five arms of the animal. Starfish, the most mobile of the echinoderms, are voracious carnivores. Their radiological images are very different from those of the echinoids, and in their great variety they remind us of a marvelous embroidery made by the hand of Nature herself.

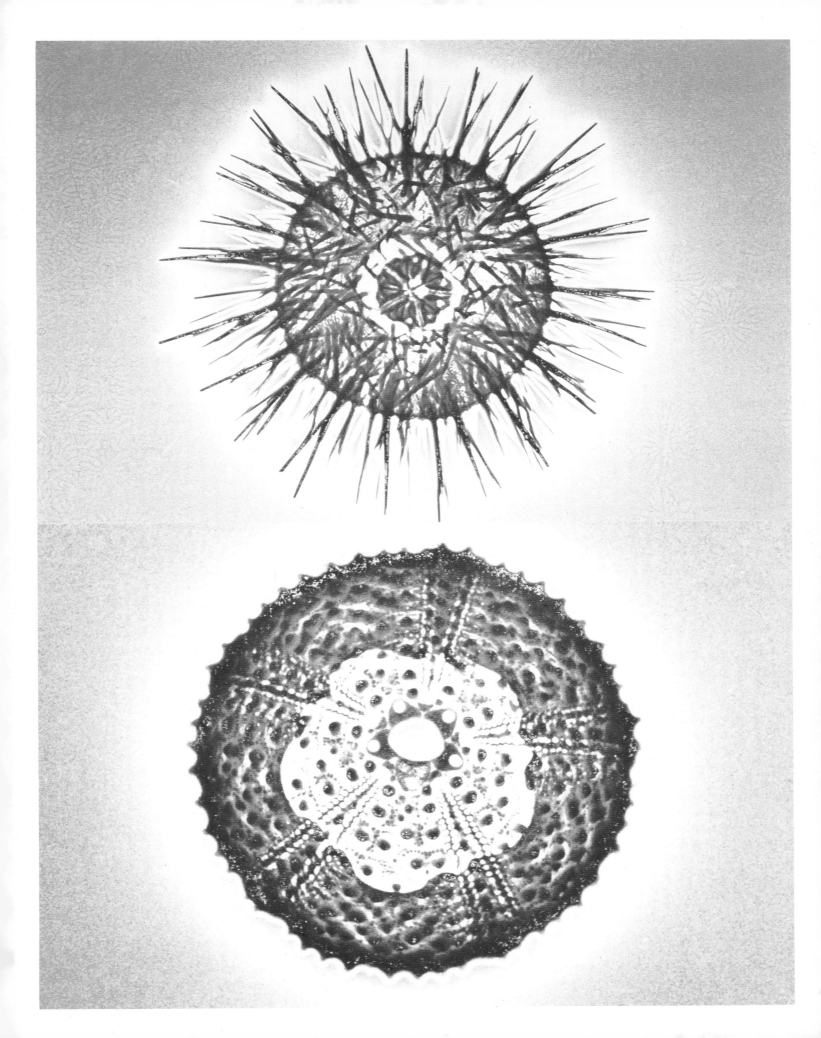

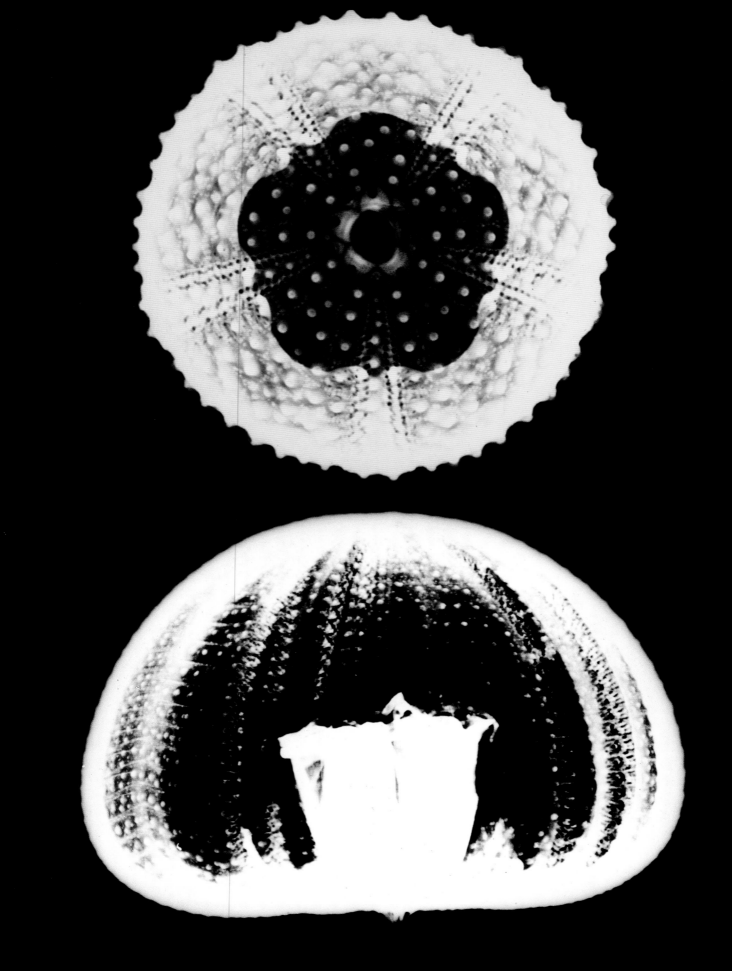

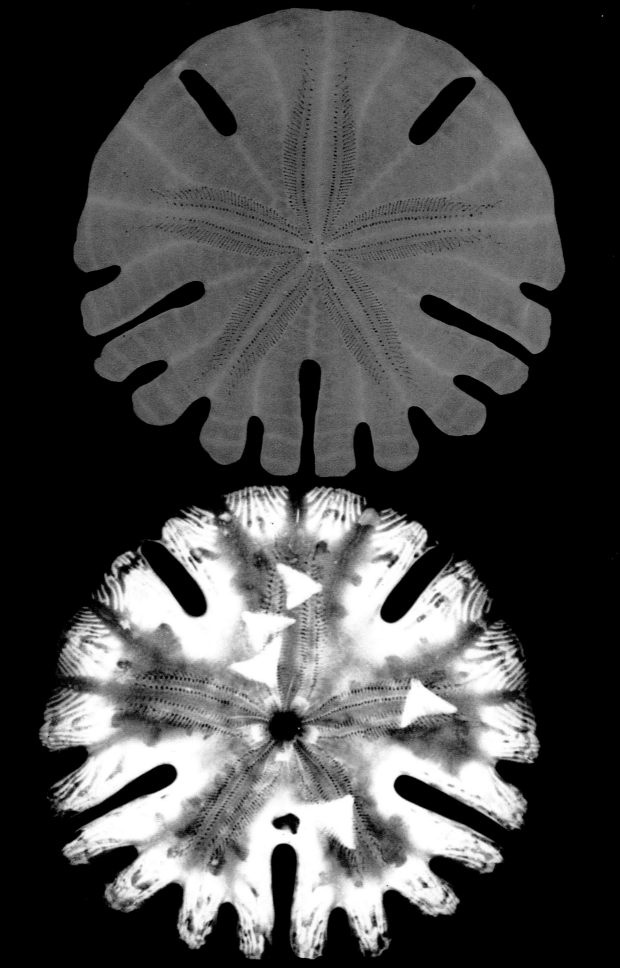

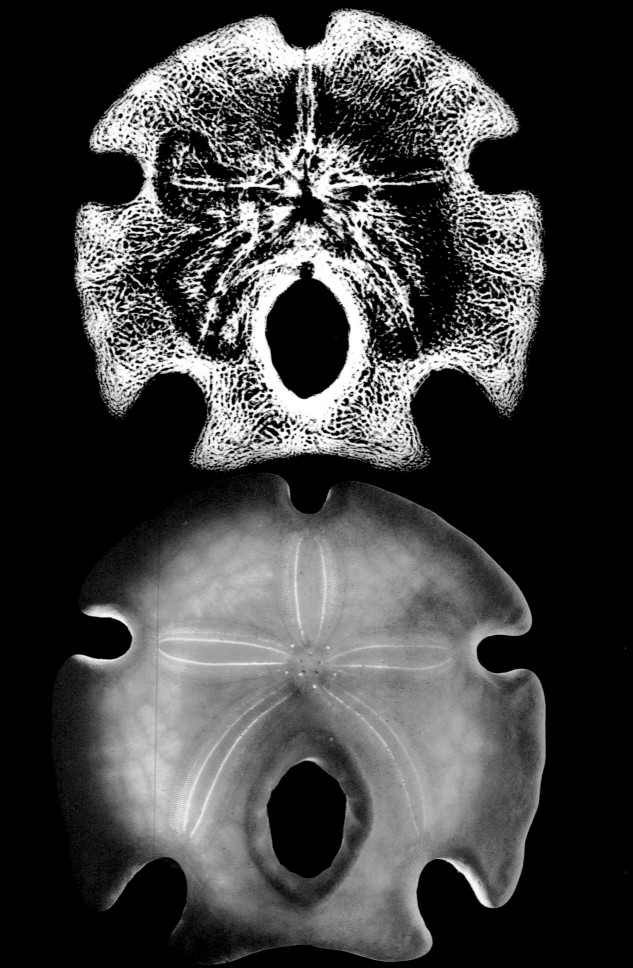

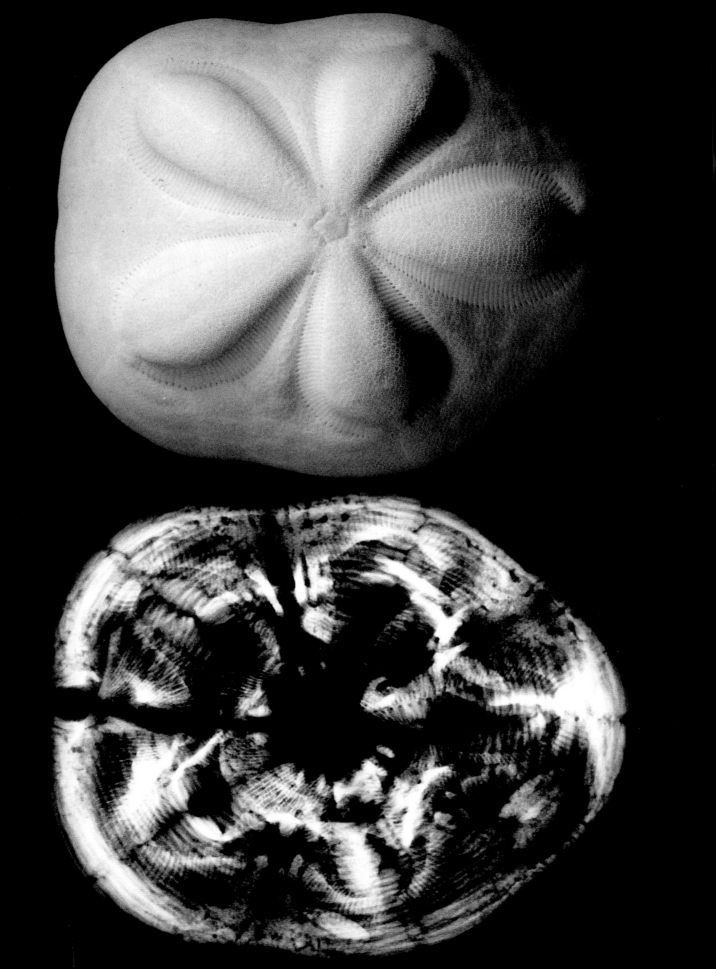

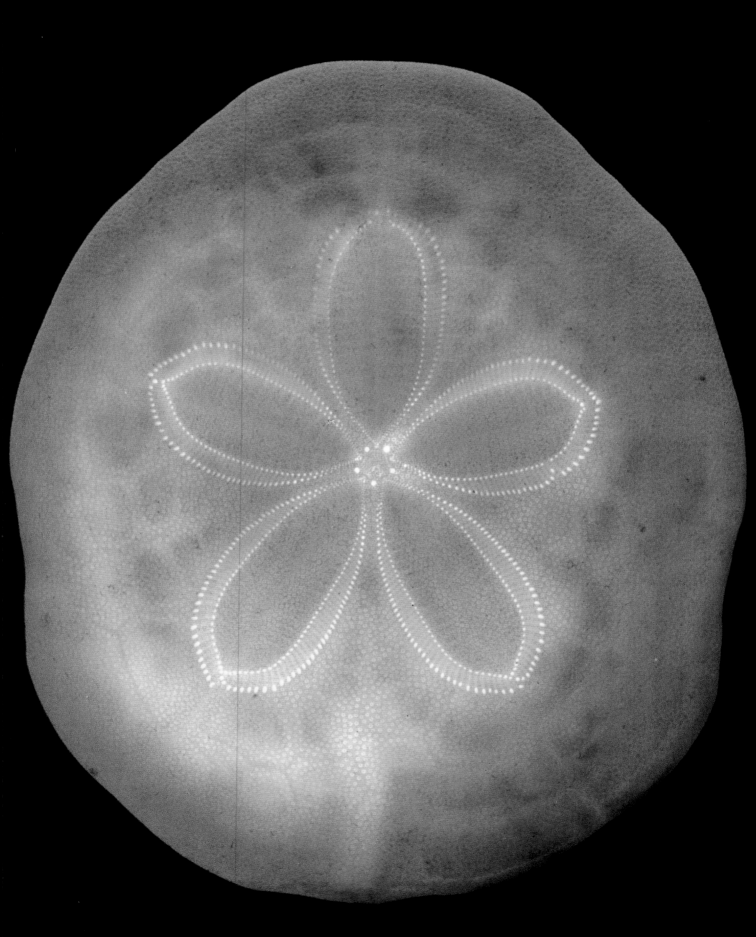

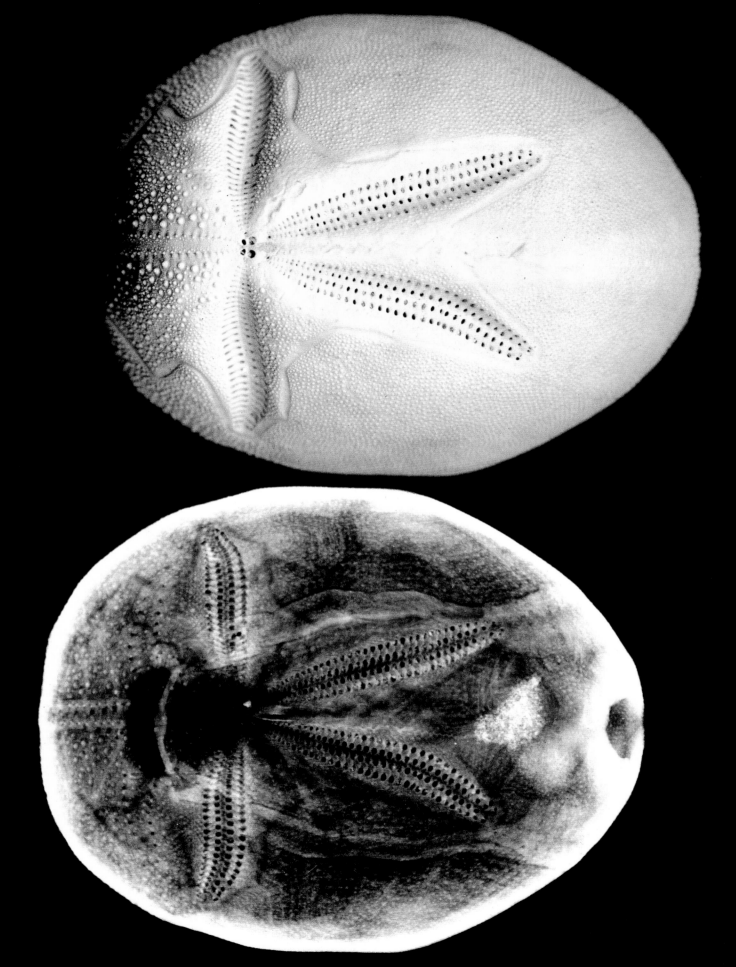

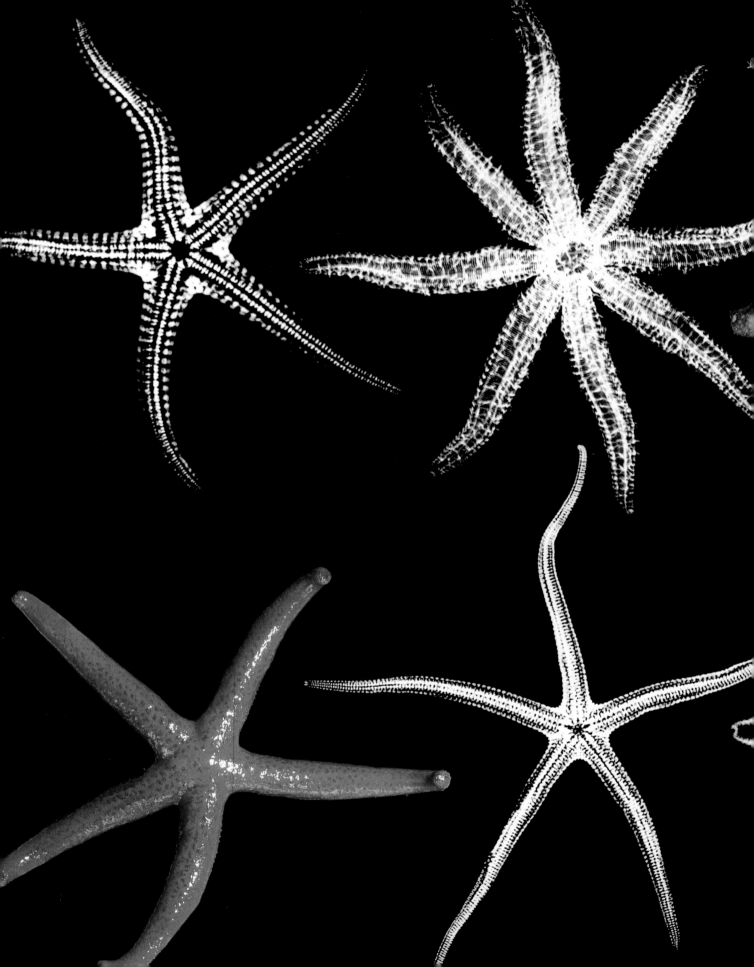

Protreaster lincki. Positive xeroradiograph. Although echinoderms are typically characterized by five-part symmetry, it is often possible to find starfish having more than five arms. These organisms will regenerate new arms in response to an injury or, sometimes, for no apparent reason. This specimen has six arms.

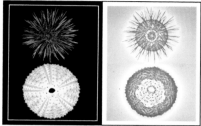

Spiny Sea Urchin (Paracentrotus lividus). Above: Photograph and positive xeroradiograph of the living animal in an axial projection. The mouth, visible at the center of the X-ray image of the living animal, shows five triangular, sturdy teeth. Below: Photograph and positive xeroradiograph of the animal's skeleton.

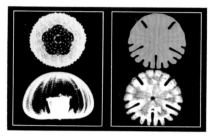

Above left: Spiny Sea Urchin (Paracentrotus lividus). Radiograph of the skeleton, printed in sepia. Below left: Sphaerechinus granularis. Radiograph of the animal from the side. The central form is the animal's chewing apparatus, known as Aristotle's lantern.

Right: Sand Dollar (Rotula orbiculus). Photograph and radiograph of the skeleton. Inside the skeleton five triangular bodies can be seen radiologically, the disarticulated teeth of the animal's chewing apparatus.

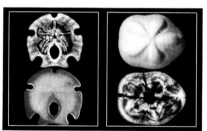

Left: Five-holed Sand Dollar (Mellita testudinata). Photograph and radiograph of the skeleton. The delicate interlaced pattern inside the skeleton of this deep-water animal is revealed radiologically.

Right: Cake Urchin (Mellita enflata). Photograph and radiograph. The skeleton of this echinoid appears extremely rich in limy formations arranged concentrically. The impressions of the internal organs are less evident in this animal than in many other irregular echinoids.

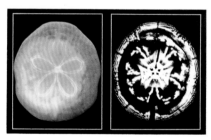

Cake Urchin (Echinodiscus sp.). Photograph and radiograph of the skeleton. The extremely delicate, transparent, petal-shaped pattern of the radial areas, or ambulacra, lit from behind in the photograph, is complemented in the radiograph by the pentagonal chewing apparatus, the animal's five teeth visible in the center.

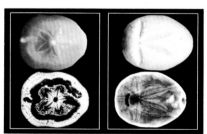

Left: Echinarachnius parma. Photograph and radiograph of the skeleton. X rays demonstrate that most of the internal surface of this thin echinoid is occupied by the intestines, which leave an evident impression inside the skeleton.

Right: Purple Spatangus (Spatangus purpureus). Photograph and radiograph of the skeleton.

Left: Composition of photographic and radiographic images of eight starfish, including Coscinasterias tenuispina, *eight-armed starfish;* Ophiciaster ophiciana, *red-colored starfish;* Astropecten bispinosus *(see below right for detail).*

Above right: Starfish (Asterina gibbosa). Radiograph.
Below right: Starfish (Astropecten bispinosus). Radiograph, detail. With X rays, the arrangement of small limy plates that make up the skeleton of the animal becomes very clear.

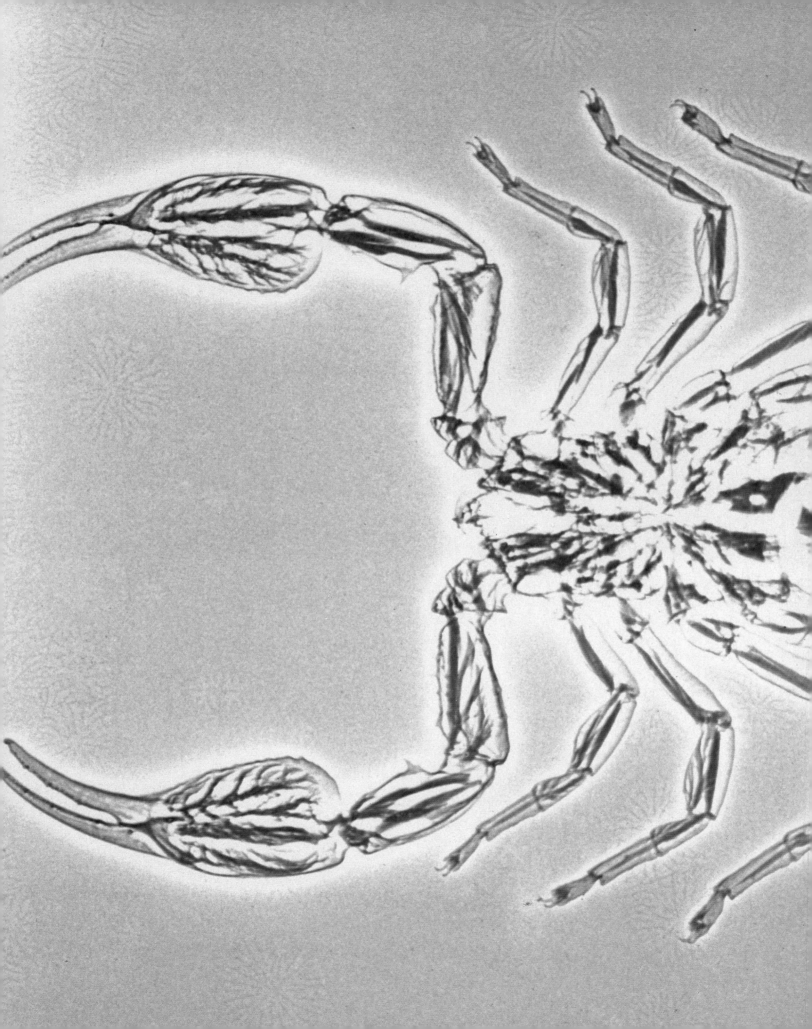

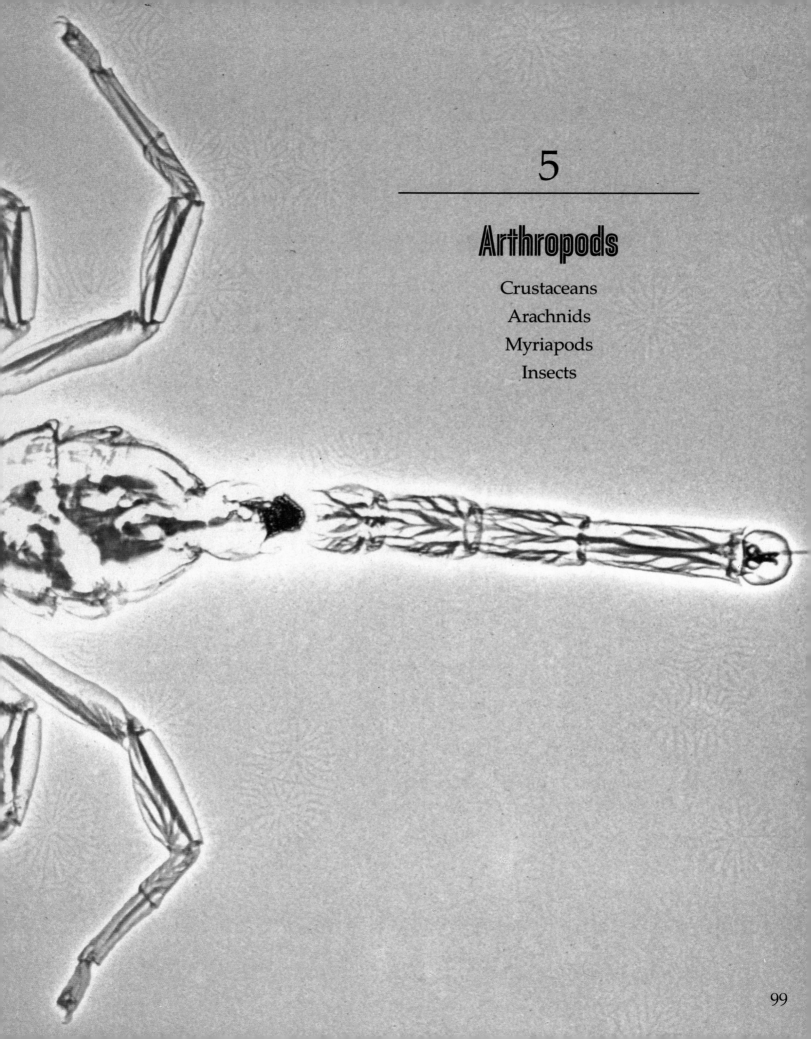

5

Arthropods

Crustaceans
Arachnids
Myriapods
Insects

From the ecological point of view, arthropods are, without doubt, the most successful animal phylum. As their name implies, they are characterized by articulated, jointed limbs, and with these they have been able to interact most successfully with their environments.

It is generally believed that the phylum had its origins in some ancestral forms of annelids (see Ch. 2), of which, incidentally, arthropods still maintain the typical segmented structure. But in general arthropods show a tendency toward unifying the segments; in their body three regions are distinguishable: head, thorax, and abdomen. A radiological investigation sometimes permits us to highlight the ancestral segmental structure in each of these regions that is not apparent externally.

An essential characteristic of arthropods is the presence of a chitin, or horny exoskeleton, having both protective and supporting functions. Thanks to this exoskeleton, radiological images of arthropods can be obtained, though with a degree of difficulty.

Four main classes of arthropods are recognized here: crustaceans, arachnids, myriapods, and insects. Crustaceans are characterized by the presence of a very resistant exoskeleton, because the chitin has been mineralized by the presence of calcium salts; the illustrative aspects of radiological images of crustaceans are therefore influenced by the frequency of the X rays used, which can be softer or harder (lower or higher).

The class *Arachnida* comprises animals such as scorpions and spiders, whose radiological images give the same sense of aggressiveness that these predators display in real life.

The class *Myriapoda* has kept closest in some ways to the annelids, the ancestors of the arthropods, and it is characterized by numerous unspecialized metameres, most of which possess a pair of walking legs. An excellent example of the class is the thousand-legger or millipede, whose segmental and repetitive structure is highlighted in the radiological image.

Insects, the widest class of arthropods and in some ways the largest animal group extant, are thus extremely successful organisms. Their success as a class must be connected with the evolution of certain innovative characteristics in them compared to other classes of arthropods, and wings are without doubt the most important of these innovations. The acquisition of wings represents an enigma in the evolutionary history of these animals: it is not possible to trace them back to an origin in the ancestral structures of the insects' progenitors; no structure in the organization of any other arthropods can be compared with wings, nor could wings have been derived from any. Among insects, wings have reached their highest expression with butterflies. The radiological image of a butterfly reveals the fantastic pattern of veins in the wing-bearing structure, often partially obscured by the presence of large amounts of pigments.

The life cycle of insects is characterized by metamorphosis, a phenomenon marking the passage from the larval stage to adulthood. A radiological image that synthesizes this phenomenon is of the cocoon of the silkworm, with the remains of the integument of the chrysalis still visible inside. The radiographic image of a honeycomb has been included among those of insects because it is the product of a very socially evolved insect: the honeybee. In its extraordinary geometric perfection the honeycomb is surely one of the symbols most emblematic of the harmony that exists in nature.

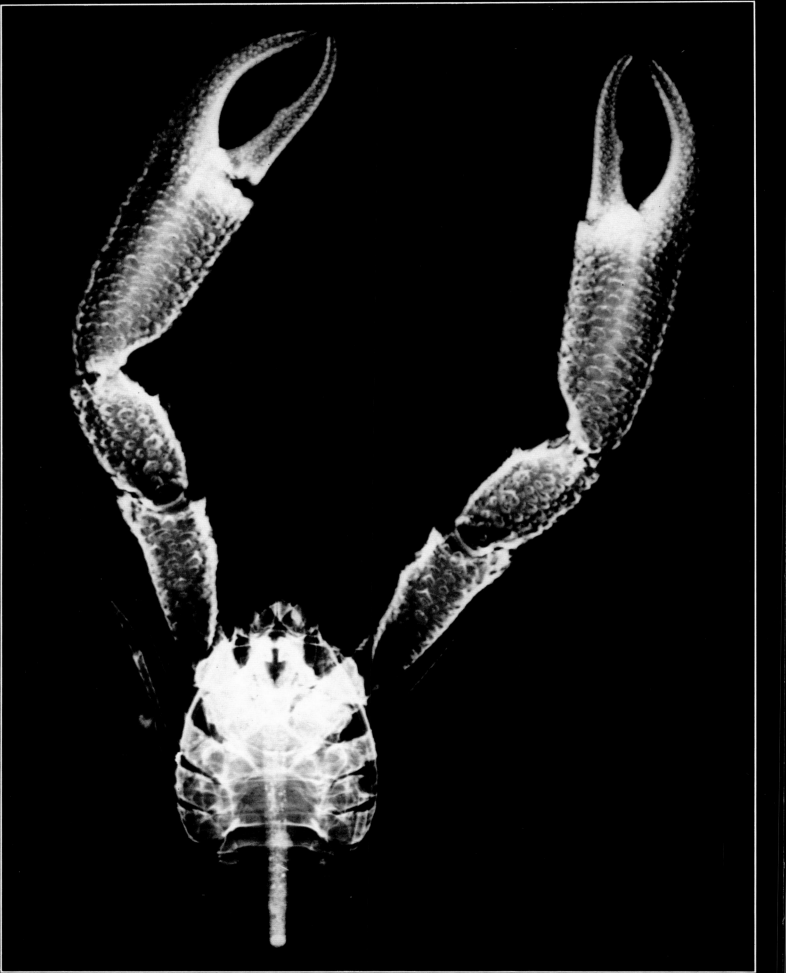

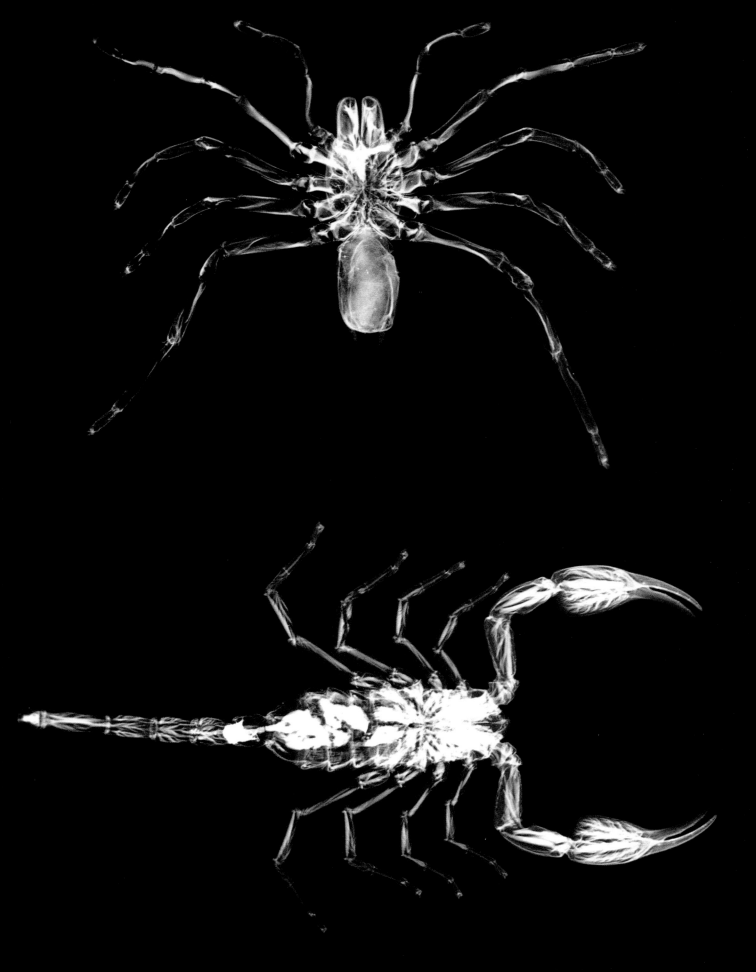

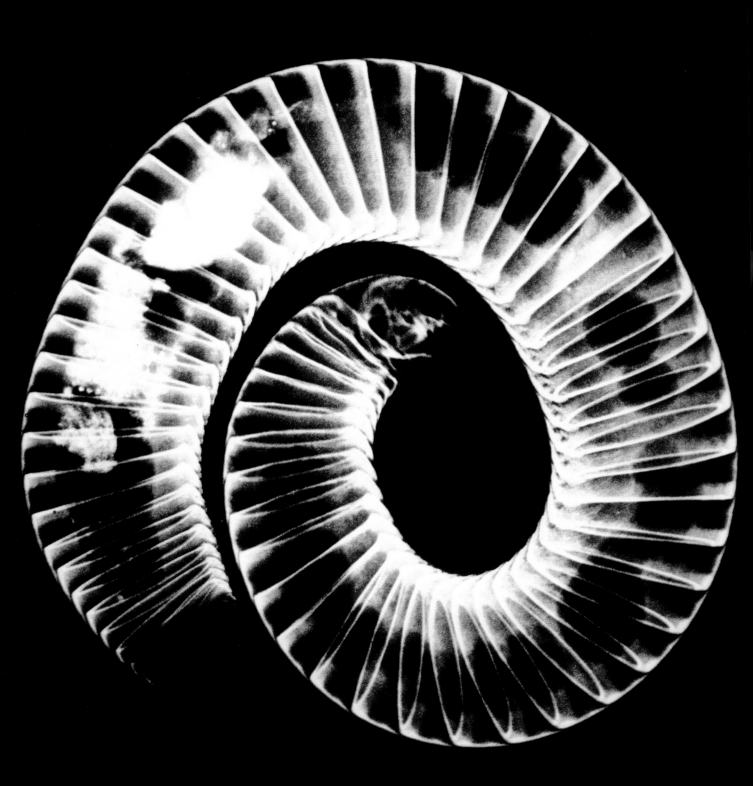

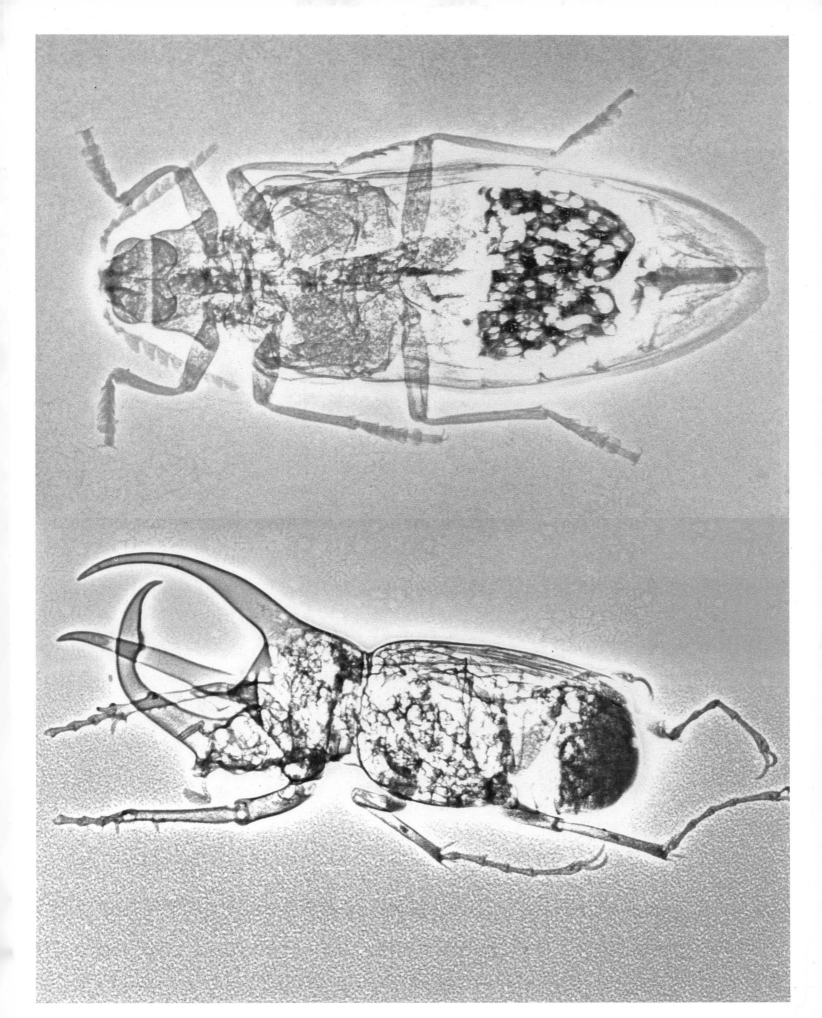

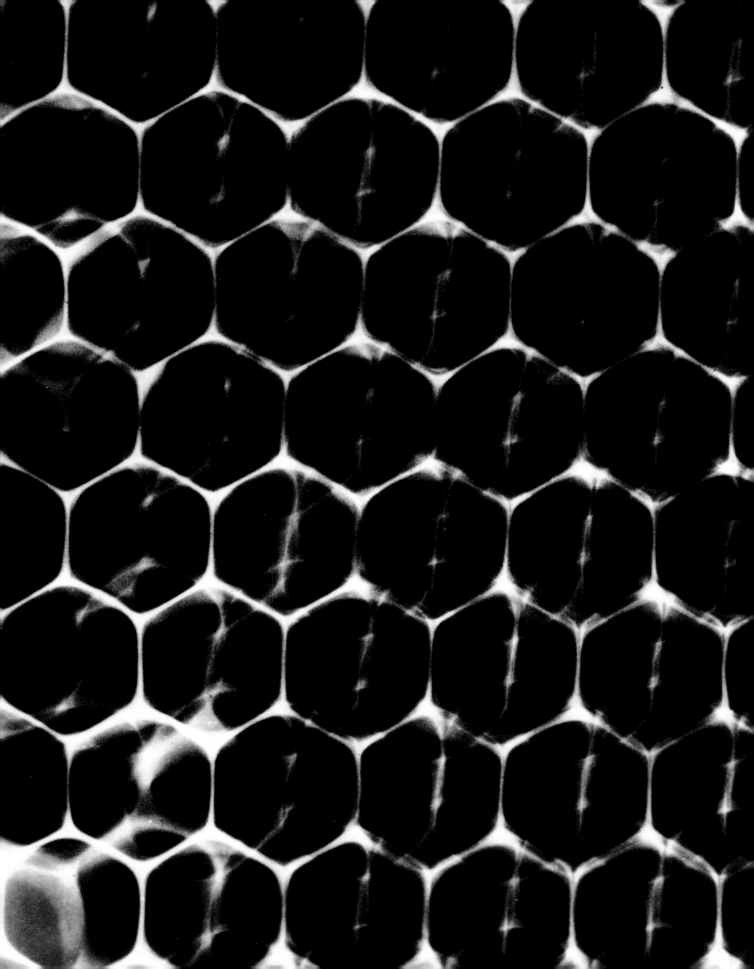

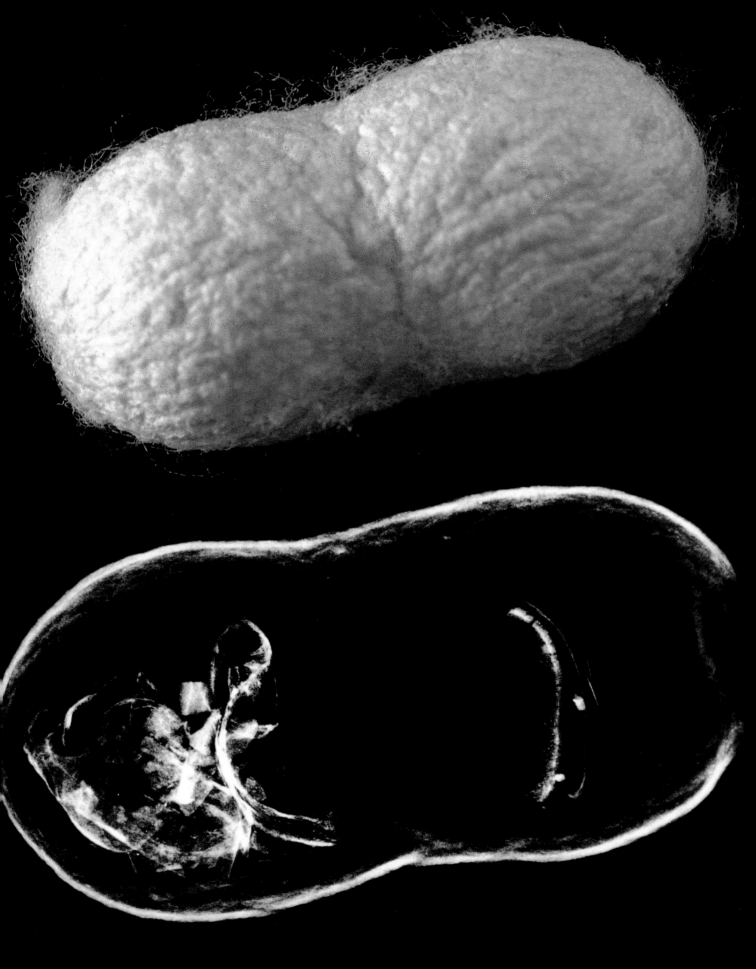

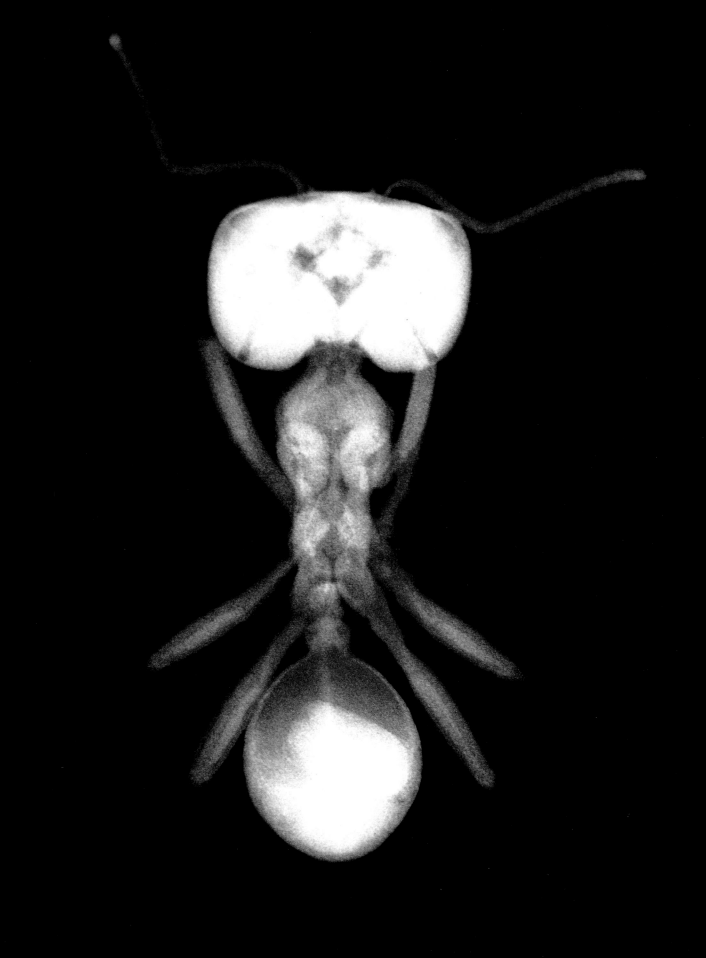

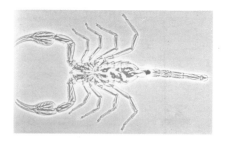

Scorpion (Pandinus sp.). Positive xeroradiograph. The radiological image of a scorpion reveals the complexity of the muscular system inside the exoskeleton of the limbs. Visible in the abdomen is some material in the process of being digested at the moment of death.

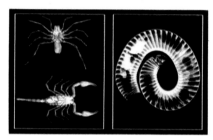

Crustaceans. Left: Lambrus sp. *Radiograph.*
Right: Box Crab (Calappa granulata). Positive xeroradiograph.

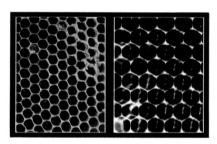

Arachnids. Above left: Spider (Migale sp.). Radiograph. The radiological image of this large spider shows us in great detail the articulation of the body of the animal and its limbs. *Below left: Scorpion (Pandinus sp.). Radiograph.*

Myriapods. Right: Thousand-legger (Iulus sp.). Radiograph. The segmented and repetitive structure of the millipede is highlighted by radiology. Food particles in the process of being digested can be observed inside the intestine of the animal.

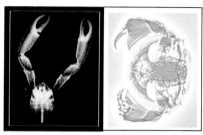

Left: Beetle (Cothaxanta opulenta). Photograph.

Above right: Beetle (Cothaxanta opulenta). Positive xeroradiograph.
Below right: Calcasoma atlas. *Positive xeroradiograph.*

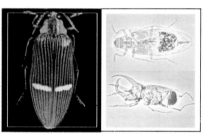

Honeybee (Apis mellifera). Photograph and radiograph of honeycombs. The radiological image was obtained by placing the honeycomb at an angle, thus creating an interplay of the hexagonal profiles of the honeycomb cells.

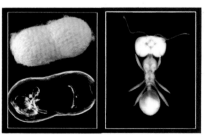

Left: Silkworm cocoon (Bombyx mori). Photograph and radiograph. Inside the cocoon can be seen the integument left behind by the chrysalis.

Right: Ant (Formica rufa). Radiograph, much enlarged. We can clearly see radiologically the mouth of the ant in the cephalic region, the joints between body and limbs in the thorax, and the intestinal contents of the abdomen.

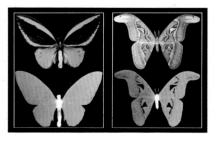

Butterflies. Left: Ornithoptera hecuba. *Photograph and radiograph.*

Center: Attacus atlas. *Photograph and radiograph.*

Right: Morpho sp. *Photograph and radiograph.*
The radiological investigation of butterflies, despite certain technical difficulties, reveals the pattern of veins that make up the bearing structure of the wings.

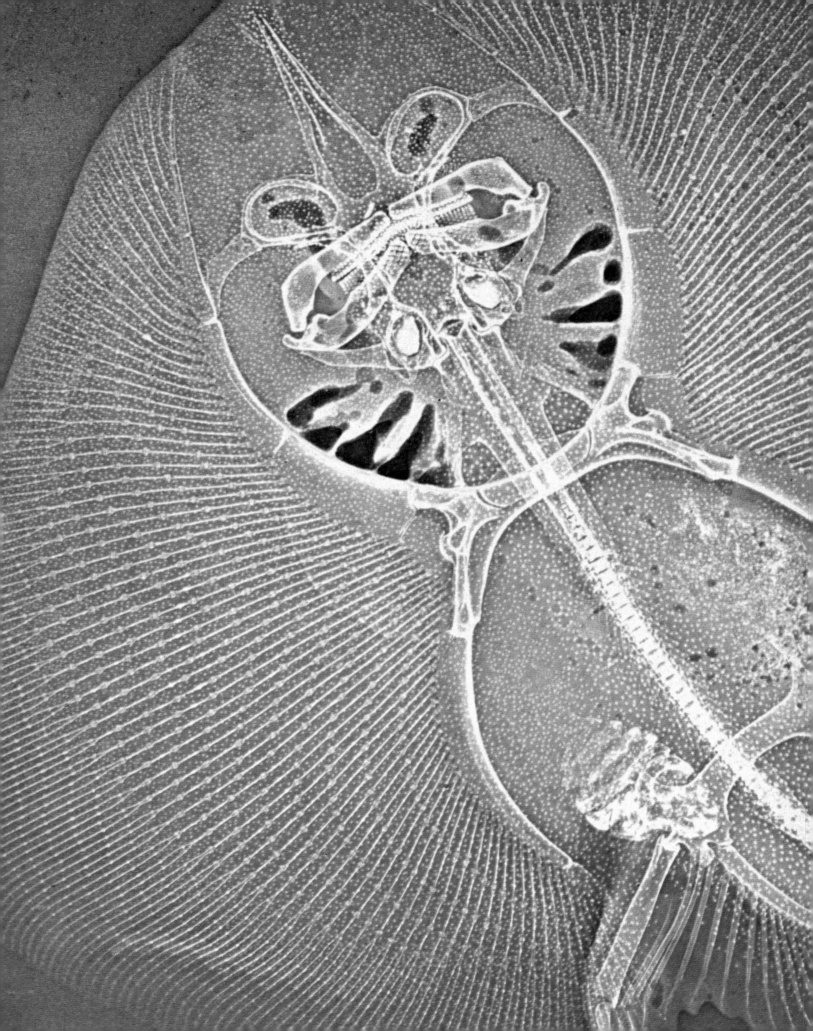

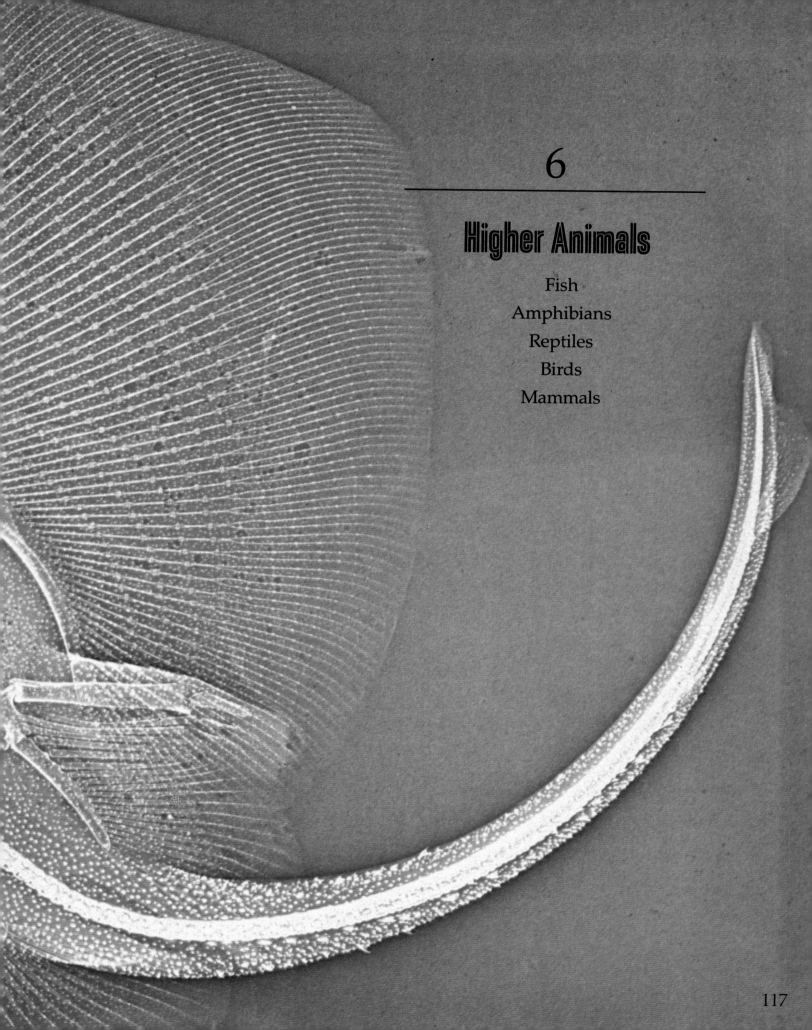

6

Higher Animals

Fish
Amphibians
Reptiles
Birds
Mammals

The first man-made distinction in classifying animals was that between vertebrates and invertebrates.

Vertebrates are, in effect, the most evolved forms of animal life. Since humans belong to this group of organisms, such a distinction might appear anthropocentric, but it emphasizes the outstanding element in vertebrates: the presence of an internal skeleton whose fundamental structure is the backbone.

As bilaterally symmetrical animals, vertebrates are classified within the phylum *Chordata*, together with a few invertebrates which share important characteristics and, in all probability, were their ancestors. These invertebrates possess an organ of great evolutionary importance, known as the notochord, a longitudinal, flexible rod of cells representing their skeleton. During the early developmental stages of vertebrates too, this is present, being later transformed into a backbone.

There are five classes of vertebrates: fish, amphibians, reptiles, birds, and mammals.

Fish are in some respects the most primitive class. The great majority of modern fish have a well-defined horny skeleton and are therefore regarded as rather evolved organisms. The skeleton of primitive fish is cartilaginous, however, and it is maintained in their most direct descendants, rays and sharks. The skeletal structures of fish are incredibly varied, and the radiological images in this book give only an idea of the numerous adaptations that fish have undergone in the course of evolution.

Amphibians, phylogenetically intermediate between fish and reptiles, are a link between the aquatic and terrestrial environments. The very name amphibian (from the Greek amphi-bios, double life) reflects the existence of two different life stages for these animals: in the larval stage, the animal lives in water and breathes through its gills, and in the adult stage the animal develops lungs that allow it to breathe and live outside water. One sequence of radiological images in this chapter shows four of the developmental stages of a frog: during the metamorphosis from tadpole to frog, we can radiologically observe the appearance of bones, first in the head, then in the back, and finally in the limbs, ending with the harmonious skeletal structure of the adult animal.

The vertebrates' complete "conquest" of land takes place with the reptiles, well represented in paleontology. Reptiles can be considered one of the most fundamental stages in the evolution of vertebrates. Dinosaurs, the reptiles' descendants, dominated the earth totally uncontested for a long period; their sudden disappearance from the scene is one of the great enigmas of science.

In the reptile class there are many morphological variations, from the *Chelona*, the armored reptiles, such as turtles, to the *Ophidia*, snakes, and the *Sauria*, lizards. Snakes, however, give us the most visually exciting radiological images, as radiographs show the infinite variety of positions they can assume.

Birds originated from reptiles, and they still share certain anatomical characteristics. The great success of birds, evidenced by their wide distribution, is certainly linked to their ability to fly. Their anatomical structure, characterized by bones lightened by air spaces, is perfectly adapted to this capability, although a number of bird species abandoned flying and readapted to an exclusively terrestrial life.

It was the hard-shelled eggs of birds and reptiles that made these animals independent of the aquatic environment in their reproduction. The shell protects the egg from dessication, its greatest danger on dry land. The radiological image of an infertile egg allows us to observe its interesting internal structure; the image of the fertile egg a few hours before hatching shows us the chick's position inside the egg.

Because the anatomical features of mammals are so familiar to us, a radiological investigation of their images seems less interesting. Bats, however, the only mammals capable of active flight, are an exception, thanks to the extraordinary development of their upper limbs.

We should like to conclude by recalling that humans (the most evolved of the vertebrates), after the discovery of X rays, have been using them for many years to study the anatomy, the physiology, and above all the pathology of our own organism. This present book is, as far as we know, the first radiological survey ever made of the living world around us.

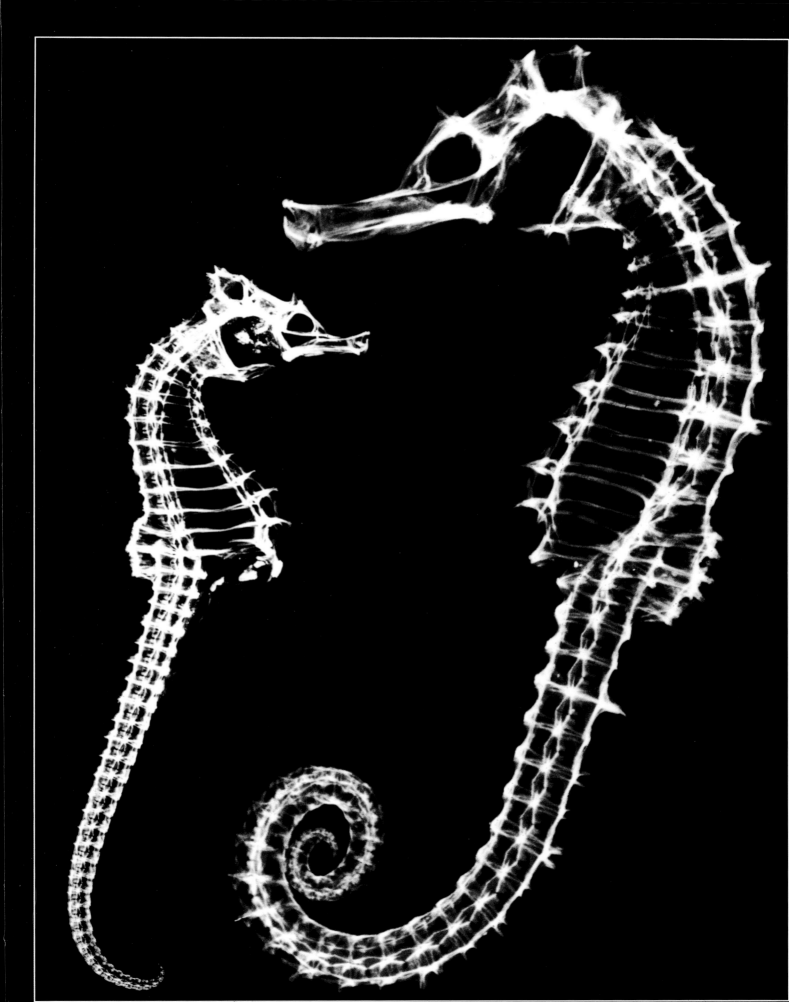

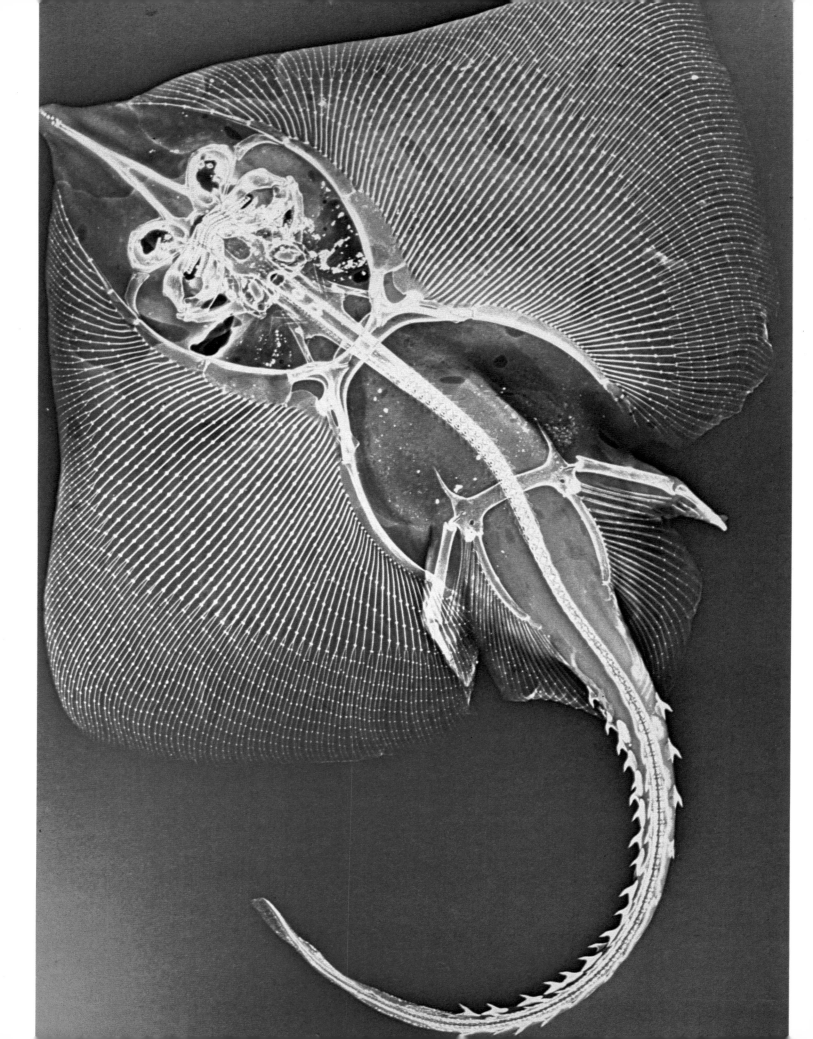

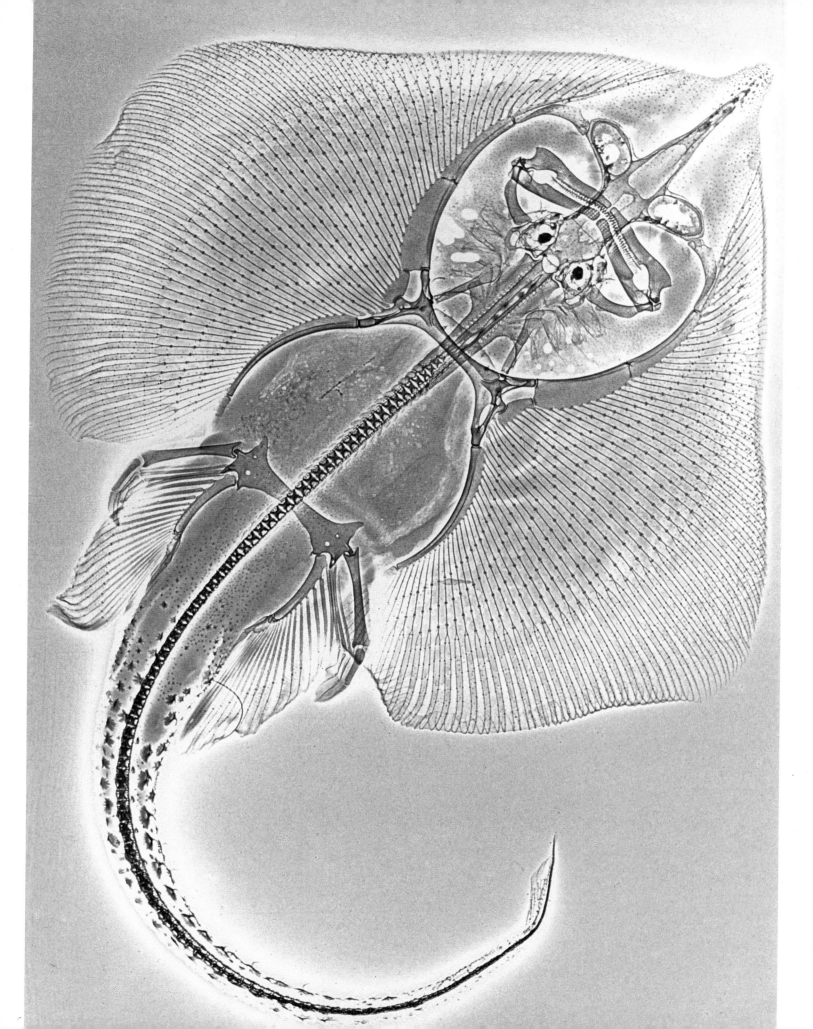

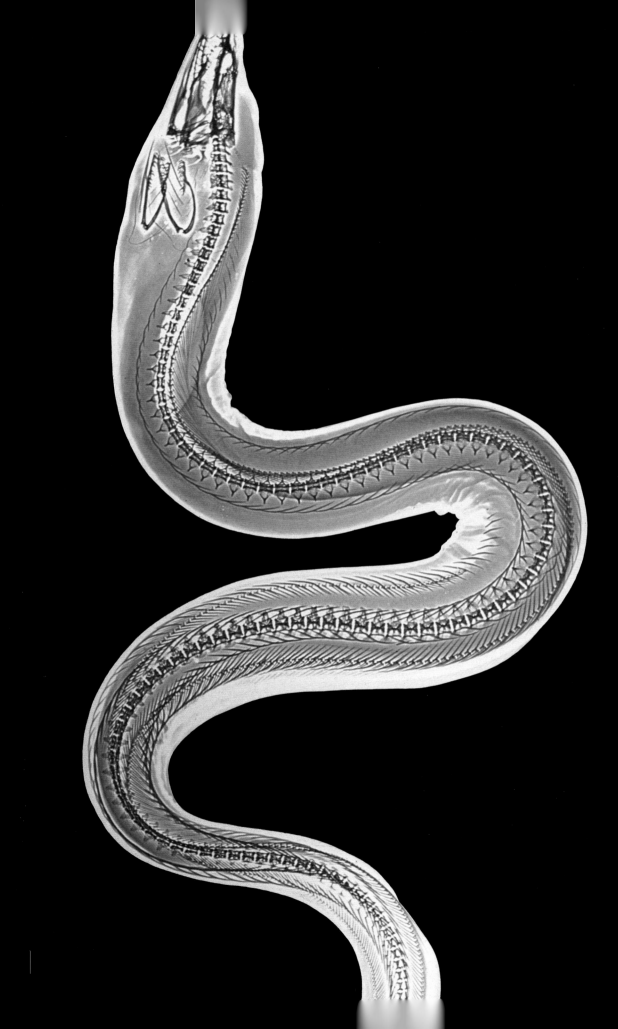

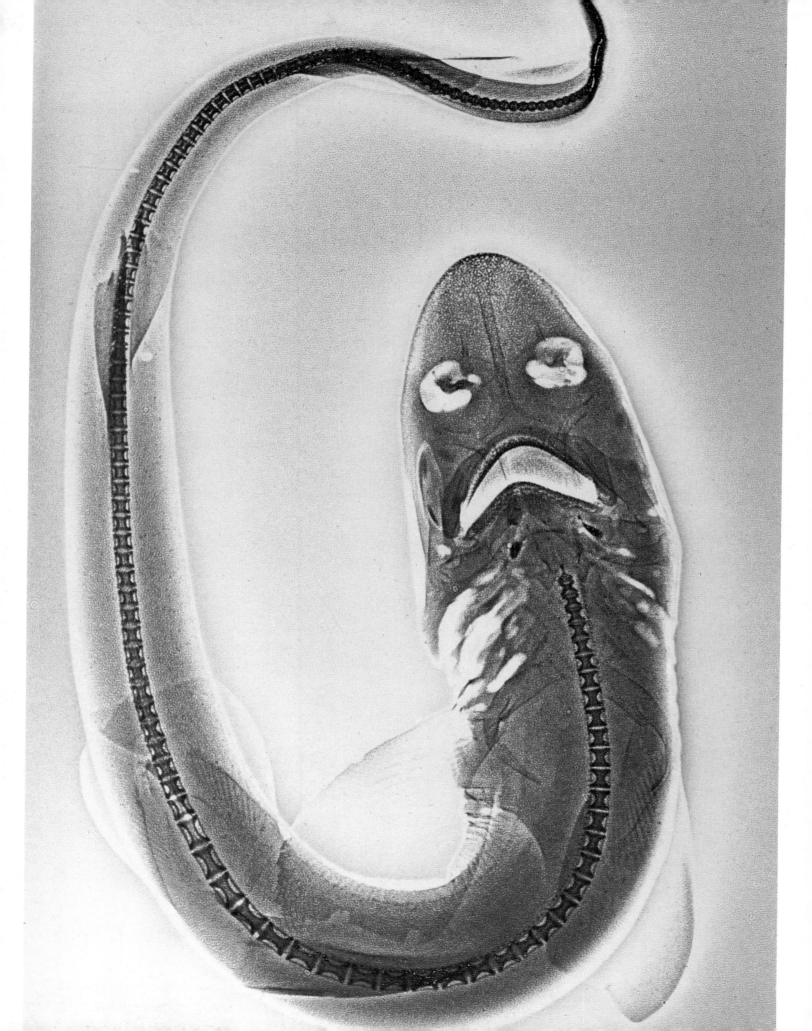

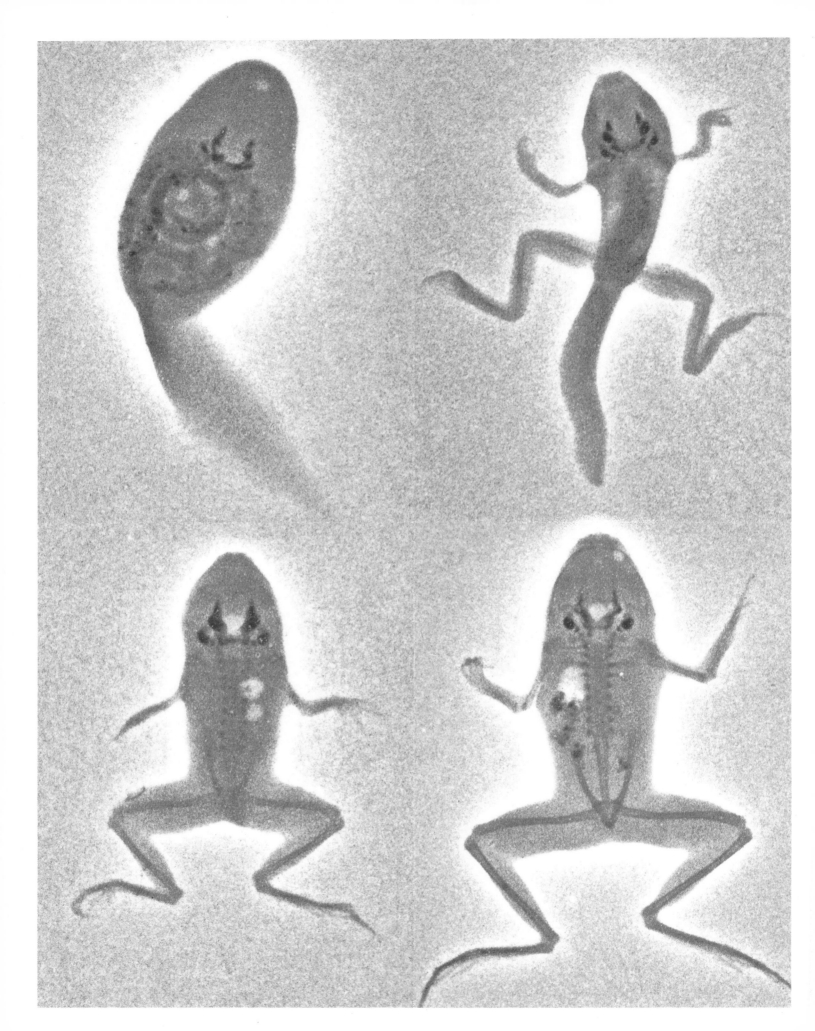

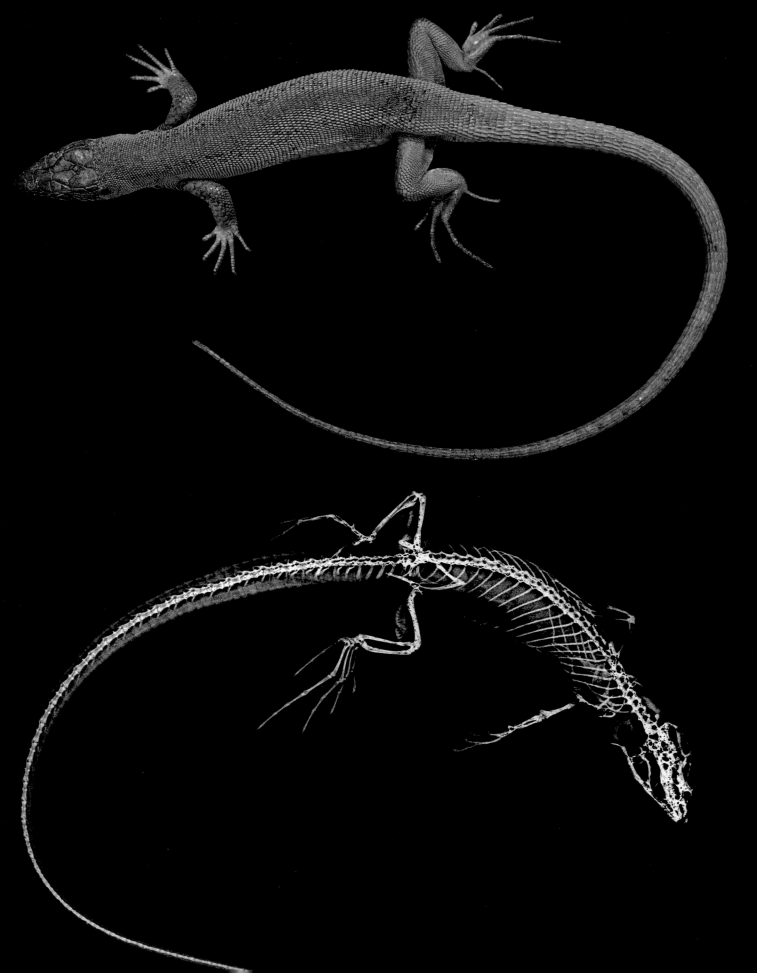

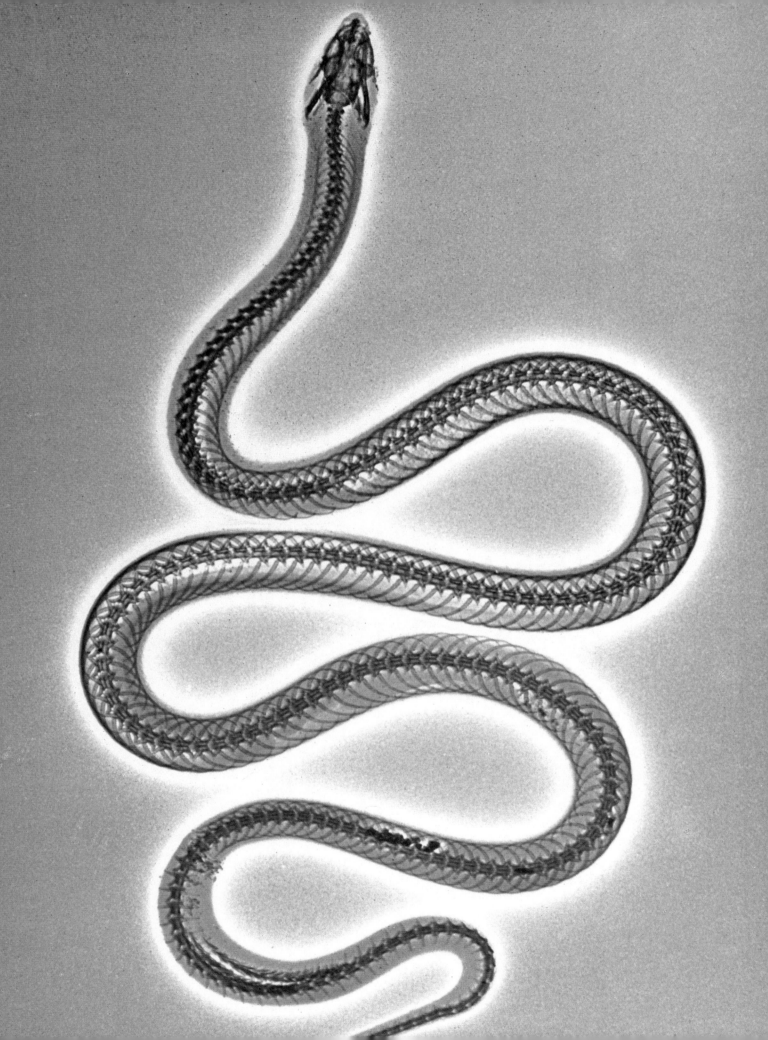

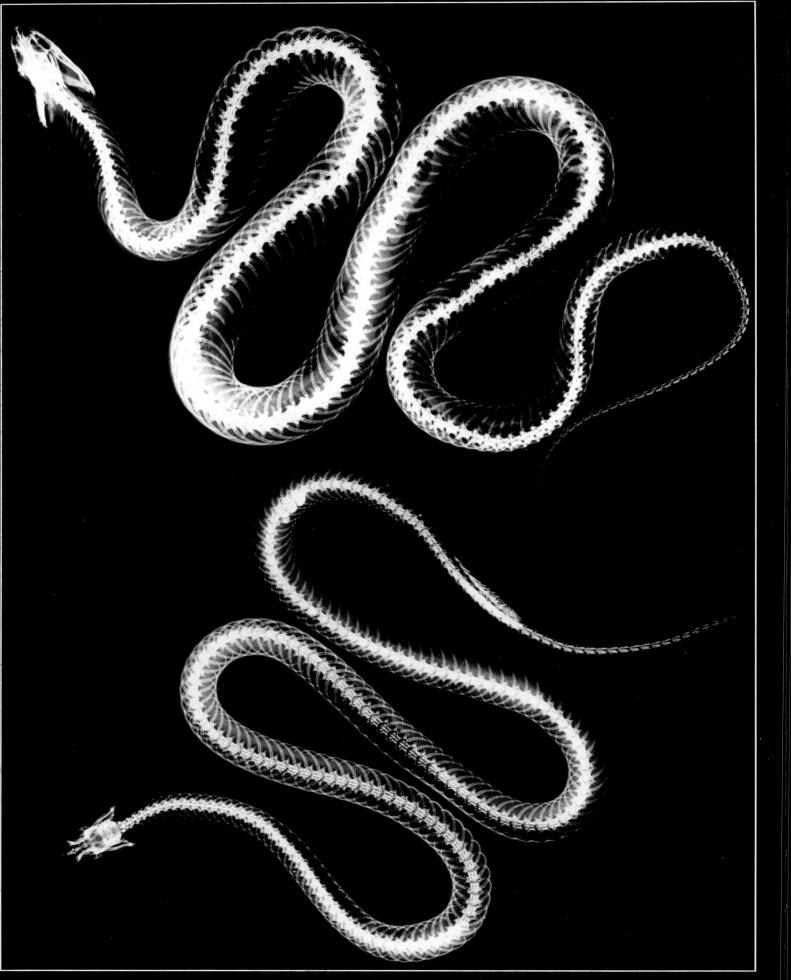

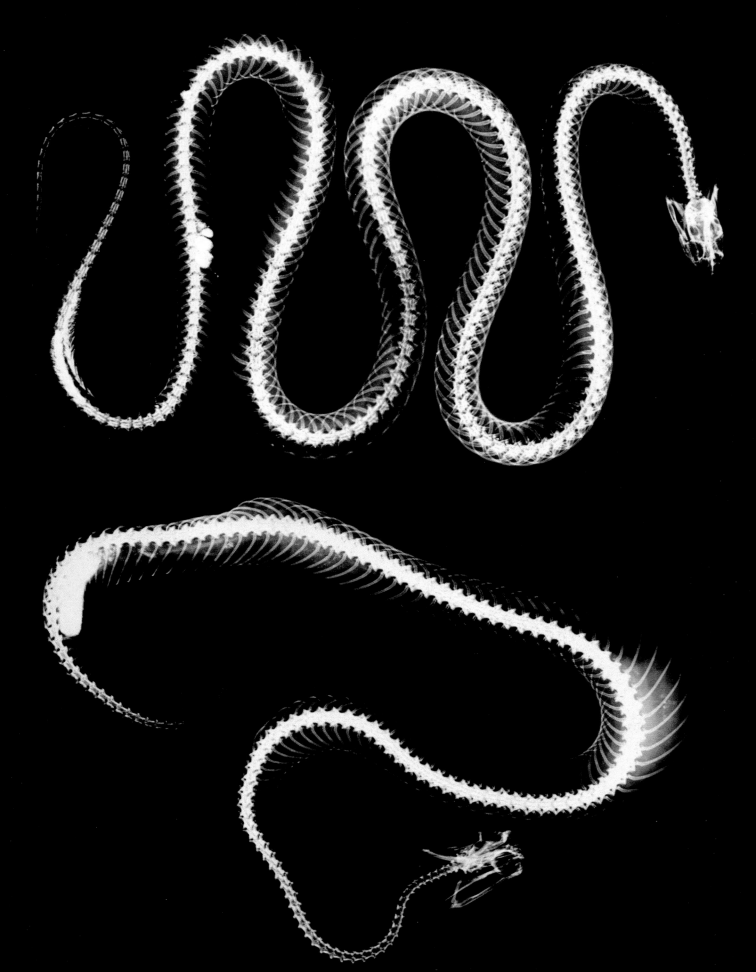

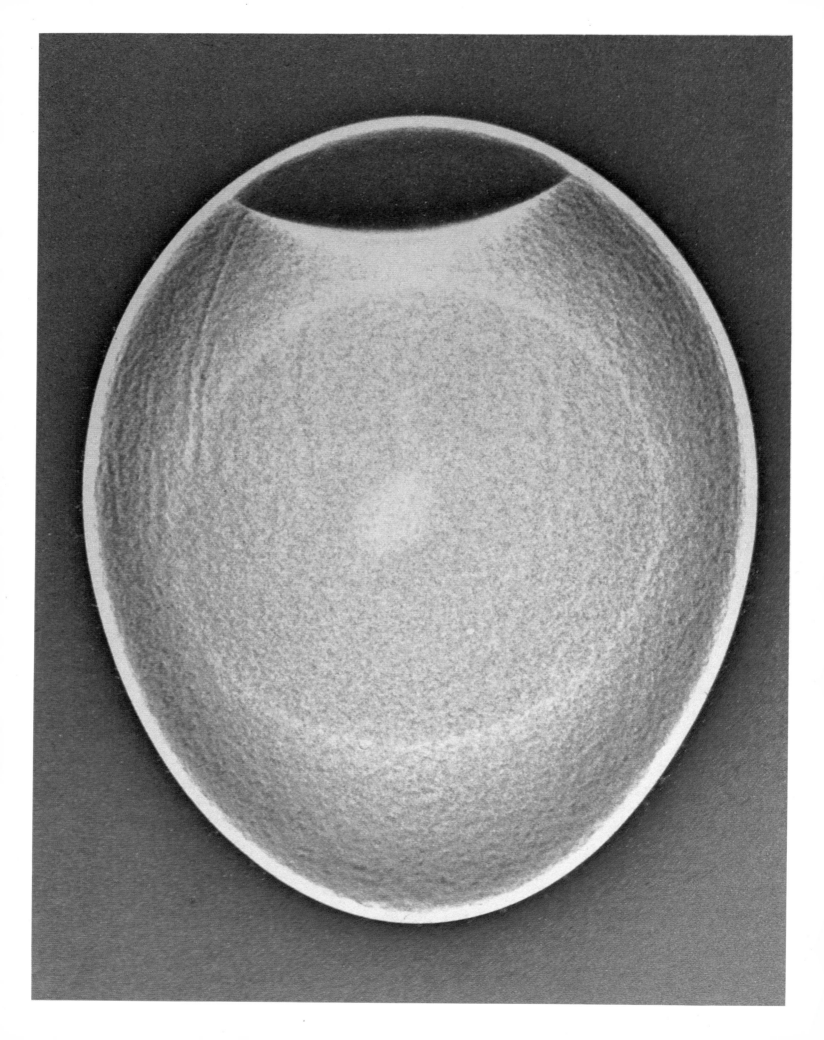

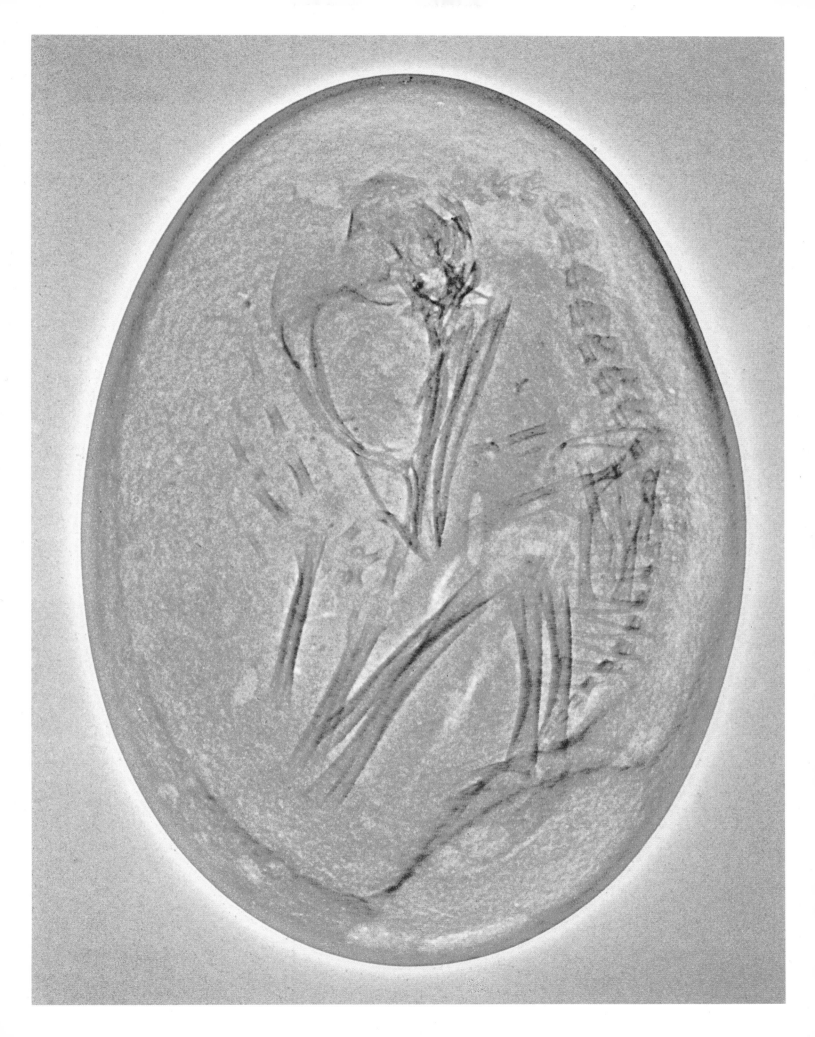

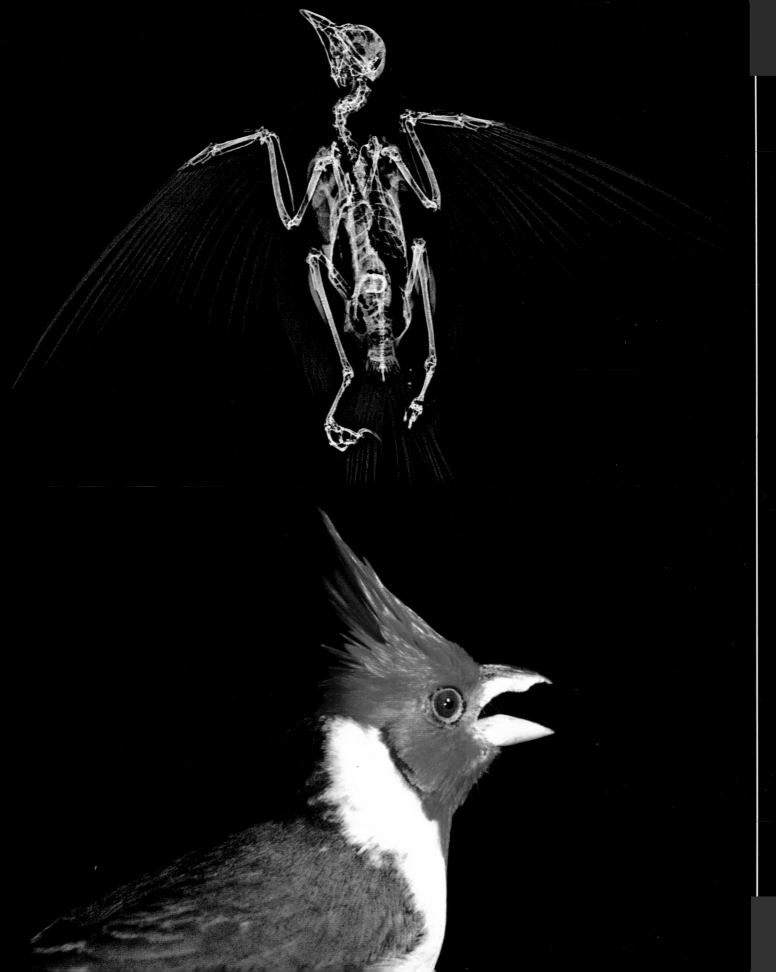

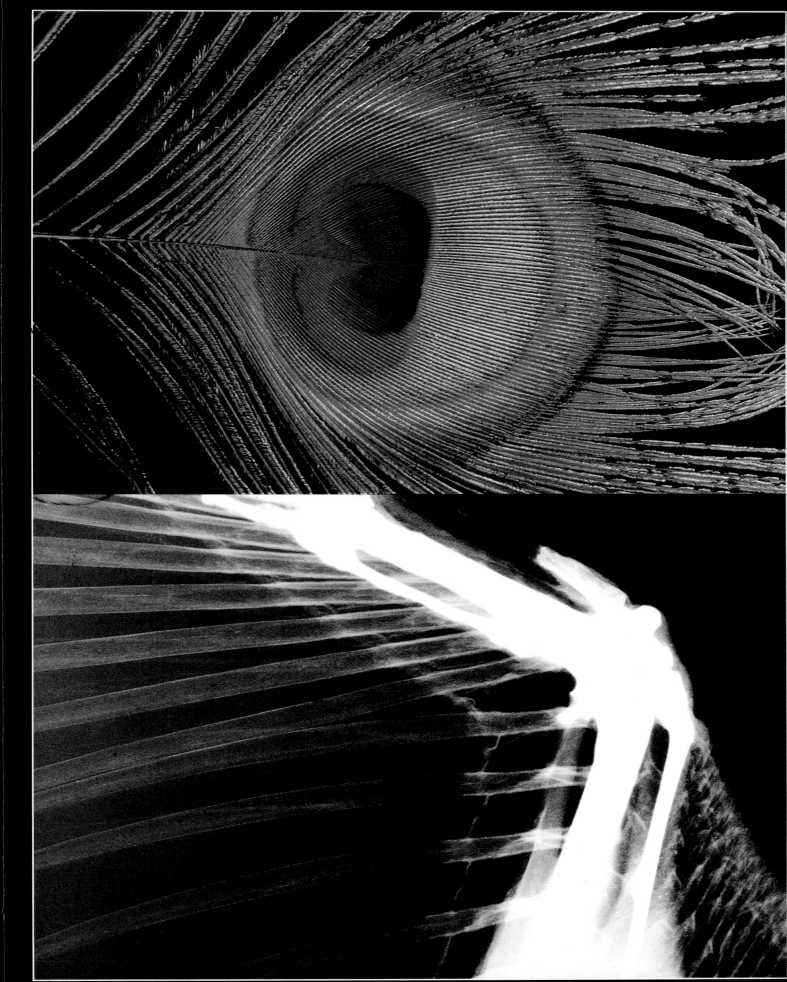

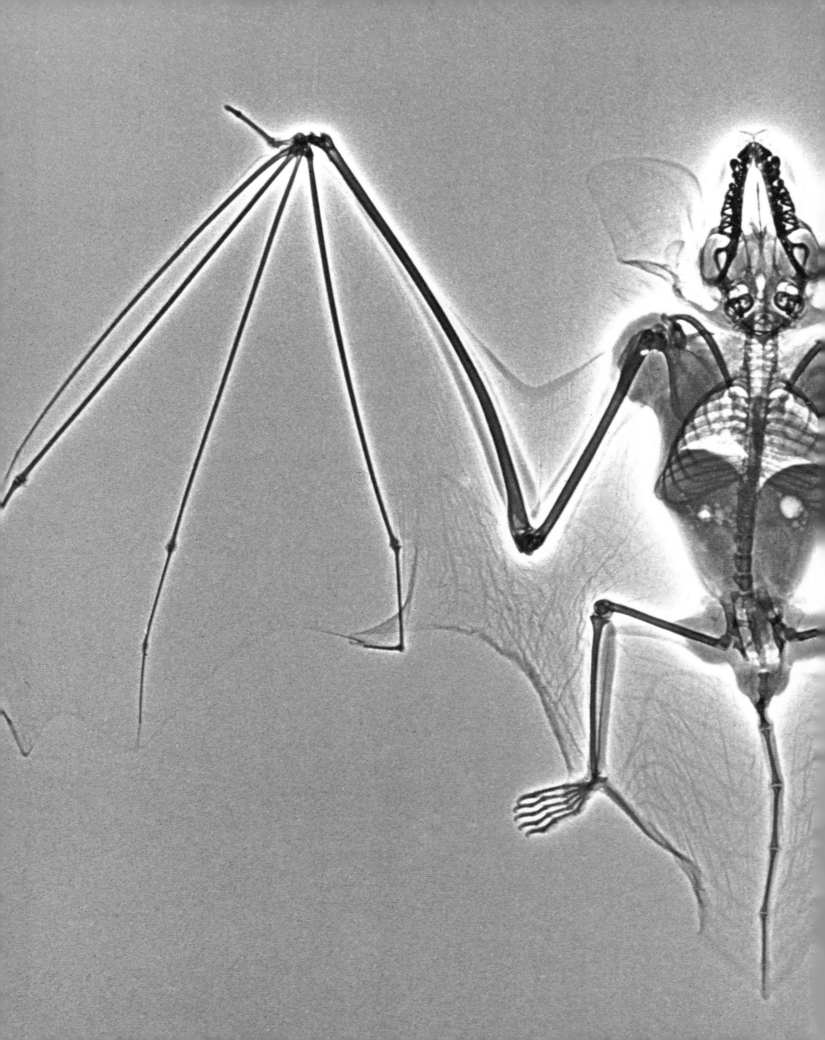

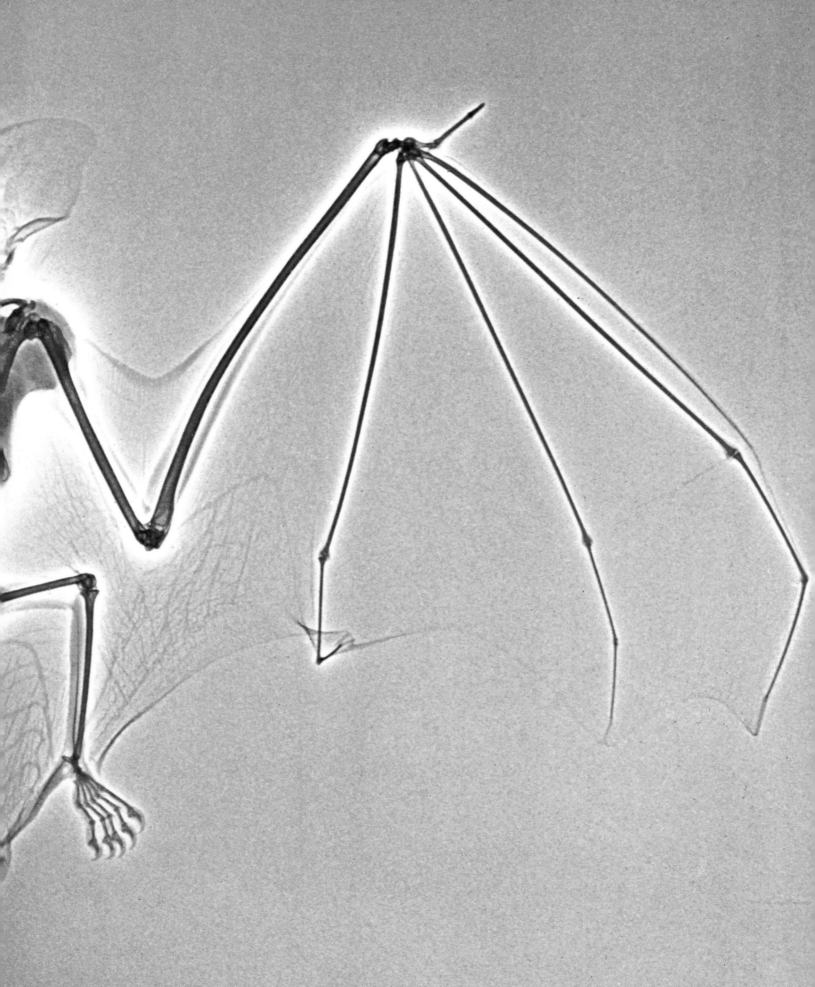

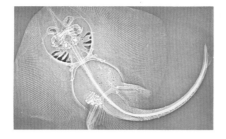

Stingray (Raja asterias). Negative xeroradiograph. Xeroradiography reveals the complexity of the ray's skeleton in its finest details. It is noteworthy that the chewing apparatus of the fish, characterized by two horizontal sturdy jaws, is capable of crushing the shells of the mollusks they feed upon. These calcareous remains have accumulated in the intestine of the animal, and are clearly visible radiologically.

Sea Horse (Hippocampus guttulatus). Radiographic images. The X rays of these two sea horses reveal the unusual structure of the skeleton. The typical vertical position of the sea horse is unlike that of any other fish species.

Chromis chromis. Positive xeroradiograph. Salient in this radiological image are the air bladders of these small freshwater fish, made visible radiologically by the gases they contain. The air bladder permits the fish to float at a given level.

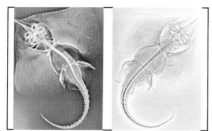

Left: Stingray (Raja fullonica). Negative xeroradiograph. The skeletal structure of the ray's large fins consists of an array of numerous cartilaginous bones, which radiologically produce a beautiful pattern.

Right: Stingray (Raja miraletus). Positive xeroradiograph. Xeroradiography in the positive mode affords a very clear view of the single backbone segments of the ray, which are less visible in the negative mode.

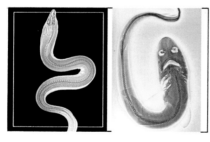

Left: Moray Eel (Muraena helena). Positive xeroradiograph. The skeletal structure of the moray eel, apparently so delicate and slender, is perfectly adapted to the aggressive life habits of this predator.

Right: Shark (Mustelus vulgaris). Positive xeroradiograph. The radiological image of this small shark synthesizes the anatomical features of this vast group of fish. Xeroradiography highlights the cartilaginous backbone particularly well.

Edible Frog (Rana esculenta). Left: Positive xeroradiographs of four stages in the metamorphosis of a frog. The images are equally enlarged. This sequence allows us to observe the progressive development of the skeleton during metamorphosis, from the limbless tadpole to the skeletally articulated frog.

Right: Positive xeroradiograph of adult frog. Less enlarged than the four opposite images.

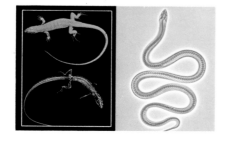

Left: Green Lizard (Lacerta viridis). Photograph and negative xeroradiograph.

Right: Rat Snake (Coronella sp.). Positive xeroradiograph. This image shows that snakes have lost their limbs completely, since even the rudiments are impossible to visualize; at the same time we can appreciate the ribs, present along almost the whole length of the backbone.

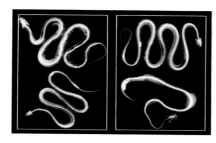

Radiographs of snakes. Above left: English Grass Snake (Natrix natrix). Some bony fragments among the skeletal components of this snake belong in all probability to a frog the snake has eaten. Below left and above right: Rat Snake (Coronella sp.).
Below right: Viper (Cerastes cerastes). This species, which lives in the Sahara Desert, has been radiographed in its most typical posture.

Left: Infertile Hen's egg (Gallus gallus). Negative xeroradiograph. Through xeroradiography the various components of the egg can be distinguished: the external calcareous shell; the air sac at top; the albumen; and the yolk, with a lighter area at the center, the germinal vessel, representing the nucleus.

Right: Fertile Hen's egg (Gallus gallus). Positive xeroradiograph showing the chick inside the egg, a few hours before hatching.

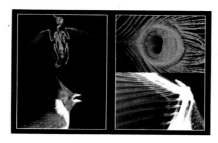

Left: Red-crested (Brazilian) Cardinal (Paroaria coronata). Negative xeroradiograph and photograph, detail.

Above right: Peacock (Pavo cristatus). Photograph of tail feather.
Below right: Radiographic detail of a bird's wing. We see the arrangement of the quills in relation to the wing bones.

Bat (Myotis myotis). Positive xeroradiograph. The overall skeletal structure of bats displays the basic anatomical characteristics of mammals. The main peculiarity of the bat's skeleton concerns the extraordinary development of the forelimbs, and especially the exceptional length of the digital bones. A cutaneous membrane stretched across the skeletal components of limbs and tail makes bats the only mammals capable of active flight.

Technical Note

It may be useful to give a brief account of certain technical procedures employed to obtain some of the radiological images in this book.

Many images have been made by xeroradiography, a process differing from traditional radiography in the manner of producing the latent image, not on film, but on an electrostatically charged selenium plate. The X rays that pass through the object being examined trigger a discharge on the selenium plate proportional to the amount of radiation that has hit the plate at each point. This latent image is converted into a real one by distributing on the plate a blue powder, the "toner," that adheres to the charged parts of the plate. The powder image is then transferred to paper and fused to it by heating.

Simply by reversing the polarity of the selenium plate, xeroradiography can yield either a positive or a negative image, and under certain conditions either of these produces images more detailed and informative than those obtained with traditional radiography. For visual reasons, some of these xeroradiographic images have been printed in black and white; others have been silhouetted against a black background.

This book contains radiological images of very small or extremely thin organisms, and to obtain them special X-ray equipment had to be used that was capable of producing extremely low-energy X rays. The high resolving power required for these images was obtained by using an X-ray tube with a very fine focus, together with fine-grain X-ray film. Of course, several technical expedients were necessary to make the most of the small degree of natural contrast in the structures of these organisms.

Great care was also taken in choosing the photographic materials used in reproducing these radiological images.